This book is dedicated to Sir Brian Baker's daughters,
Jean and Margaret (Margot),
and all his grandchildren.

Also by Jacquie Buttriss

A Muddy Trench: A Sniper's Bullet (Pen & Sword)

The Great War Ace,
The Red Baron and Beyond

The Great War Ace, The Red Baron and Beyond

The Life and Achievements of Air Marshal Sir Brian Baker

KBE, CB, DSO, MC, AFC

Jacquie Buttriss

Pen & Sword
AVIATION

First published in Great Britain in 2024 by
Pen & Sword Aviation
An imprint of Pen & Sword Books Limited
Yorkshire – Philadelphia

ISBN 978 1 39905 831 5

A CIP catalogue record for this book is
available from the British Library

Typeset by Mac Style
Printed in the UK by CPI Group (UK) Ltd, Croydon, CR0 4YY.

Pen & Sword Books Limited incorporates the imprints of After
the Battle, Atlas, Archaeology, Aviation, Discovery, Family History,
Fiction, History, Maritime, Military, Military Classics, Politics,
Select, Transport, True Crime, Air World, Frontline Publishing, Leo
Cooper, Remember When, Seaforth Publishing, The Praetorian Press,
Wharncliffe Local History, Wharncliffe Transport, Wharncliffe True
Crime and White Owl.

For a complete list of Pen & Sword titles please contact

PEN & SWORD BOOKS LIMITED
47 Church Street, Barnsley, South Yorkshire, S70 2AS, England
E-mail: enquiries@pen-and-sword.co.uk
Website: www.pen-and-sword.co.uk
or
PEN AND SWORD BOOKS
1950 Lawrence Road, Havertown, PA 19083, USA
E-mail: uspen-and-sword@casematepublishers.com
Website: www.penandswordbooks.com

Contents

Foreword

I have so many happy memories of spending time with my grandfather as my twin brother and I were growing up. Every summer we would go and stay with him and Granny at St Andrews for two weeks. As the years passed, I have found out more and more about my grandfather's incredible career in the Royal Flying Corps and Royal Air Force. As we grew older, it became apparent that his story had to be told, for the sake of our children, the wider public and generations to come.

My great-aunt Winnie, Sir Brian's sister, amazingly kept many of the letters he wrote to her during the First World War and after. She also kept albums full of photographs, citations, invitations to the Palace and newspaper articles he wrote for the local papers of St Andrews and Hertford. There are many photographs taken by Auntie Winnie of my grandfather's RFC pilot friends at Queen's Hill House, the family home in Hertford. These were discovered when the house was cleared, following her death, some years after her brother had died.

As I began to collate this material, my mother, Margot, the only surviving daughter of Sir Brian and now in her nineties, provided me with even more relevant letters and documents she had kept. She has also provided me with many memories and anecdotes from her childhood. Looking back, she talks of her pride in her father and all his achievements, but a sacrifice was made regarding family life, with the many postings overseas.

Sir Brian's nephew, Malcolm Fleming, also provided me with information that was relevant to the book, including photographs and articles. He also shared with me his personal memories of spending time with Brian during school holidays. Sadly, Malcolm died before the project was completed, but he was very excited by it and had been looking forward to reading the book.

With all this material, it became apparent that a professional writer was needed. I was drawn to Jacquie Buttriss as the author for her experience of writing such a biography. She has written a number of military books, with positive reviews from previous clients and the press. In addition, when we met, I felt we could work well together. She also gave me the confidence of knowing that she is a historian and voluntary museum researcher. I read one of her more recent books that had a similar theme to this one, in being a biography of a military officer of the First World War. His story had some parallels, but it was the ease of reading the text that gave me further confidence in commissioning her to write this book.

Jacquie has tracked down even more material and successfully woven her considerable research into the narrative, thereby adding to and enriching what we already knew of Sir Brian's life. Indeed, she has helped us to bring the character and exploits of this amazing man to the attention of a new generation.

In addition to its wider readership, this book will be of interest to military historians. Very few books seem to have included details of the development of aerial warfare, from a gallant wave to their opponent, to firing a revolver from an open cockpit, to operating a machine gun through the propeller.

Sir Brian flew through that most dangerous period of 1915 to 1918 and survived to achieve equally great success in the Second World War and beyond, including having a pivotal role in the development of aircraft carriers, as well as participating in both D-Day and the Berlin Airlift.

The stories of those who did not survive the wars live on through my grandfather's letters, articles and speeches. I hope this book does him proud.

Nigel Butler, grandson of Sir Brian Baker

Chapter One

1912–1915

Daring Feats

'These tricks are all very well, but it's madness to think that flying machines will ever be of any use.'
(Overheard by schoolboy Brian Baker at a flying display at Hendon Aerodrome in 1910. Brian was certain this man was wrong, but how could he prove it?)

B orn in 1896, Brian Baker's childhood paralleled the earliest years of aviation, and by the age of 13, he had already developed a fascination for flight. As he later wrote in his old school magazine of the people and events that had inspired him:

I was at school during a fascinating period in aviation history. [Louis] Blériot had made his historic flight across the Channel the year before I entered Lawrence House at Haileybury College in 1910. My housemaster was an authority on the flight of birds, particularly their take-offs and landings.

Every week I read *The Aeroplane* magazine from cover to cover. Its photographs of extremely daring flying feats and my visits to Hendon Aerodrome encouraged my keenness to fly.

All the famous pioneer pilots we there at Hendon – men like [Claude] Grahame-White, Philippe Marty and Lord Brabazon. They flew round and round the airfield, then raced between the pylons.

Their aircrafts were funny old things in the air, looking rather as if they were hooked up in the sky because they hardly seemed to move. Mind you, they were doing 30–40mph in Blériots, Farmans and Boxkites.

I'll never forget the excitement the day B.C. Hucks did the first loop ... and I was there to see it. He simply flipped over backwards in a tiny loop, but as the gasps and cheers demonstrated, it was a most astonishing and daring thing to do – a marvellous feat.

Watching those chaps in their flying machines, that were more or less paper and string, finally determined me to try my hand at flying when I left school.

In fact, Brian's fascination was sparked even more strongly by two events in the summer of 1913, when he was 16. Soon after the end of term, Brian was taking part in army cadet manoeuvres – the first time they included aircraft. Every morning, Brian watched enthralled as the planes flew to and fro overhead. As he later wrote: 'Exciting days they were.'

The other event could have had the opposite effect. One sunny morning in September 1913, Brian went out for a cycle ride with a friend, up the Graveley Road near Stevenage. As they meandered along, the distinctive sound of a labouring engine, about 300 feet up in the sky, caught their attention. They stopped and looked up, just in time to see a familiar shape – a Morane Deperdussin monoplane, with 'RFC' (Royal Flying Corps) painted on the fuselage. It looked unsteady, pitching up and down. Brian couldn't take his eyes off it. He later described the incident in an interview for the local newspaper: 'Suddenly, there was a loud bang, like an explosion. The engine flew to bits, went through the main spar and the wings broke off. Everything crashed to the ground.'

Brian and his pal left their bikes against the fence, climbed over and ran towards the wreckage. Brian noticed another man running across from the far side of the field and they all arrived on the scene at the same time. No doubt, Brian dreaded what might have happened to the crew. However, it was immediately evident that both the pilot and his observer had been killed on impact. A horse-drawn ambulance arrived, followed by the local bobby, and before long a small crowd had gathered. Finally, a photographer set up his camera tripod to take a photo of the debris from the crash, surrounded by the onlookers, including Brian and his friend.

Several years later, he wrote:

> I can still see the scene vividly in my mind's eye, the noise, the parts sheering off and then the silence. A memorial to the unfortunate crew of the Morane Deperdussin was placed at the roadside near the village of Willian and has remained there ever since.

In fact, the memorial stone is still there today, more than 100 years later.

Brian never forgot this traumatic event, but rather than put him off the idea of flying, it made him all the more determined to pursue his ambition, not only to fly but also to help improve the safety of aircraft.

On leaving Haileybury College in December 1914, 18-year-old Brian's parents might have hoped he would join the family business in Hertford, but he had long nurtured other ideas … and the outbreak of war had changed everything.

As an outdoors young man, talented in all sports at school, Brian was keen for action and adventure, and, unexpectedly, the war would provide just the chance he was looking for. Along with his group of friends at Haileybury, he had been an enthusiastic member of the school's Officers' Training Corps, where they learned all the basics of being a soldier and the skills of being a leader. This experience entitled Brian to enlist in the army as a second lieutenant, with all the responsibility and opportunity that would bring. Thus, he signed his papers to join the Rifle Brigade and began his army career, training with his platoon at Purfleet.

Many years later, Brian wrote another article for his old school magazine, *The Haileyburian*, looking back at the events of this time, in which he expressed:

When war broke out in 1914, I found myself with a commission in the Rifle Brigade. I was, however, very much attracted to aviation, and in my battalion, I found a kindred spirit in another officer, who was mad keen to transfer to the Royal Flying Corps. I'm afraid we were more interested in watching the flying at Joyce Green, on the other side of the river, than learning musketry at Purfleet.

Finally, the news came that the Royal Flying Corps were recruiting for pilots. Brian and his Rifle Brigade chum filled in their application forms and, after a few tense days of waiting, were called for interview. They both passed through and their transfer papers arrived the following week. Brian's orders were to travel north to Montrose Air Station in Scotland, which had been the RFC's first operational airfield in Britain – a site chosen so that it could protect the nearby Royal Navy bases and patrol the northern seas.

After a few days' leave at home, the day came in mid-September 1915 when Brian embarked on the long train journey from Hertford to Aberdeen, where an RFC mechanic's vehicle was waiting to take him on the final leg of his journey to Montrose. As they passed through the rural landscape and approached the airfield, Brian's excitement grew. Finally he caught his first sight of the vast field itself, with its huge sheds that served as aeroplane hangars for an assortment of beautiful flying machines that he had once described as having looked like they were made of paper and string, but in fact consisted mainly of strong wires and strengthened linen. There was no landing strip in those early days, just a large expanse of grass that stretched along the length of the field. Along one side, there were huts and an old farmhouse that were used as accommodation, offices and stores. However, Brian's gaze would have no doubt rested on the planes themselves.

In the first few days, Brian settled into his billet and got to know his new companions. All the eager trainee officers were required to spend their first days at Montrose learning the basics of flight and familiarising themselves with the planes on site. As we know from his letters, Brian was impatient to get into the cockpit for his first flight training session, but in the meantime he chatted to the mechanics to learn all he could about the planes and sought rare opportunities to go up on local flights as a passenger. His chance soon came, as he related in a letter to his sister Winifred:

Royal Flying Corps, Montrose
Sept. 26th 1915

My dear Win,

Many thanks for your letter and Mother's. I have had two more flights and the last one I was allowed to control the machine in the air from the front seat. It is very easy when you once know how to do it, but requires some practice afterwards to become good at it.

I was up with Maxwell, sitting in behind and had my hands on the dual controls along with his from behind me. He had his arms either side of me, when suddenly he raised his arms above his head and I knew I was driving the whole show, so I held on tight and we hovered about. Wonderful. We were about fifteen hundred feet up so it was quite safe and he corrected anything I did wrong immediately. Tell Mother I learnt to fly with a chap's arms round my waist!

It rained hard and blew awful all day yesterday, so we had nothing, and there was no early flying this morning because it was still raining. We stay up here until we have flying certificates and then we are sent off to Coastal Flying School on Salisbury Plain to finish off and get our wings.

We have to know all about the engines, construction of a plane, Lewis machine guns and signalling, as well as map-reading and map-making and weather forecasts, etc.

Most of the people here are Scottish and nearly all of them have been out to the front – a treat for many as observers. One of them actually drilled down a German aeroplane with a Lewis machine gun, when making flying reconnaissance of German lines.

I am billeted at the organist's house, so there is a continual hymn of praise going all day.

The aerodrome is about one mile from this barracks and is a very large open space, with the sea running alongside the west of it.

There are three large sheds and we have three Maurice Farmans with front elevators – the older type and are very easy to fly. Two new Maurice Farmans are without front elevators. We also have eight Curtis tractor biplanes which are just being assembled and look very nice machines and are pretty fast, doing about 70 or 80 miles per hour.

Two Martinsyde biplanes are not used for instruction, but the C.O. sometimes goes out in them. He crashed one the other day by hovering to land in a field and broke the undercarriage.

One Caudron biplane was crashed by another fellow the other day, by the engine failing when he was doing a cross-country and could not find a decent place to land. He struck a fence and did in the undercarriage but did not hurt himself.

We pupils are taken for one or two long joy-rides and these are only allowed round and round the aerodrome, until we have passed our forced landing tests, when we can do some short cross-country flights.

The view of the hills, the coast and the sea in the early morning when the mist is rising from about a thousand feet is simply a treat and the sunset just as pretty. When you are in an aeroplane, there is wonderfully little sensation. I will try to describe my first flight (with an experienced pilot). I got in the thing, just as I was, without my cap. The engine started and there was the deuce of a row behind me, with backfires, etc. This went on for about half a minute, then the thing brisked up and away we went, breezing across the ground at what seemed to be a terrific pace. All of a sudden the speed decreased to take off and the breezing stopped. The next thing I knew we were in the air and gliding along. It was a lovely sensation as there were no bangers or anything like that. The higher we went the slower we seemed to go.

The earth seemed to get onto one side of us, but we seemed to be sitting quite upright and there was no tendency to fall out or anything like that.

The only time you feel you are in the air is just as the engine is cut off to come down nose first. You just feel a dropping sensation for about fifty feet and then you don't feel anything more than a rush of wind, until you flatten out and you are bumping along the ground again.

There is absolutely nothing to be nervous about in the whole show and there is nothing in the world to compare with flying along at a good height across country. Houses look like dots, roads look like threads and the ground like a chess-board. You look far across the miles and get a lovely view of countryside and hills, right across Forfarshire.

There is plenty of fishing here. Someone went out the other day and caught thirty-six codlings in about two hours and got tired of hauling them up. There is also fishing in the sea for sea-trout. The people who have been here longest are asked to different places around to shoot, so my turn will come. One can get guns and fishing tools here, so no need to send mine up.

I won't try to come home because it costs such a deuce of a lot and I don't want to lose any flying time. I will be allowed to go up solo soon, as the hours are great, so it won't be more than a week or two.

Love to all,
Brian

With opportunities most days for a 'joy-ride' with an experienced pilot round the aerodrome and sometimes a little beyond, it was only nine days before Brian had jubilant news of his first solo flight to relate to his sister:

Royal Flying Corps, Montrose
Oct. 5th 1915

My dear Win,
Just a line to tell you that I went out alone yesterday morning for the first time, and again in the afternoon. I did a landing from 1,500 feet with the engine off. I think I've created two records for Montrose. The youngest member of the squadron and taken my solo flight in the shortest time. I've only had one and three quarters of an hour tuition, so surprised them a bit.

It has been grand weather up to date this week, but I had such an exciting experience this morning. I went up about 6.30 and when I had been round once, a thick mist suddenly descended and I couldn't see a blessed thing, so I came down to 200 feet and still could not find out where I was. After coming a little lower, I saw a large tree in a field, which I knew was next door to the aerodrome, so I shut off and down I came and landed absolutely in the rim of the aerodrome. It was very exciting, but I was mighty glad to be down again.

I shall be coming to Gosport before very long with the R.A.S., and with a certain amount of luck I may fly down as a passenger. I hope so.

Tell 'E' that very shortly I shall be efficient enough to take her up if she can manage to disguise herself as a mechanic.
Yours with love,
Brian.

On 25 October, Brian Baker took and passed his test in a Maurice Farman Longhorn biplane and was awarded his flying certificate. In his next letter, he told Win what the test had involved:

To get my ticket, I had to do ten hours of flight over a given course and land, so it was not very difficult. It would be very different doing it on a Curtis because it lands at a pretty fair pace, about

55 miles an hour. I shall get five cross-country flights now, such as Perth, St. Andrews, etc., which will be a bit of all right.

On 3 November, Brian was still at Montrose and writing to Win about pheasant shoots, goose chases and bicycle outings to Dundee, Arbroath and Edinburgh. On one of these evenings, he and his companions saw an astonishing sight:

> The night before last we had a fine aurora of Northern Lights. It was just like a crowd of search-lights shooting out a very bright show. They kept on shooting and then dying down again. Some of them reached up to the Pole Star. I wish you could have seen it.
>
> We have 2 B.E.c planes now and another coming, so things are looking up a bit. The Maurice Farmans are going with the RAS [Reserve Aeroplane Squadron]. So we shall have 2 Carter Aero B.E.2cs, Martingales and all very fast machines, all tractors.
>
> It's getting remarkable here in the wind. The gale last week was simply terrific. The sea was a wonderful sight – mountains high.

Now Brian had to do some practice cross-country flights on his own or with one other, but, as he put it:

> We were governed almost entirely by the weather conditions, since instruments were non-existent and we relied on our eyes to find the way. Forecasting was rather rough and ready. The Flight Commander would hold his hankie in the air to see how strong the wind was. If it was too much, we couldn't fly that day.

On a good day, for newly solo pilots, more experienced fliers would help them pick out a route and a clear landing place. On Brian's first short flight, he did well until it was time to land and he forgot that without his instructor, the plane was lighter, so it was higher at the mark, so he hopelessly overshot, but fortunately he landed successfully the second time.

His next cross-country was to Laurencekirk, about which he wrote:

That field holds a lasting memory. I was accompanied by a fellow officer called Seedhouse, an Olympic hurdler who was learning to fly. When we touched down, the whole of the local school turned out to welcome us. It was at the school's field and the landing of an aeroplane was still quite an event. Along with their schoolmistress, they gathered round for a chat. Soon it was time for us to fly back to Montrose. As we trundled past them on take-off, the teacher called for 'three hearty cheers for our gallant airmen'.

The childish cheers ringing in our ears, we charged along staggered into the air … and caught the trees at the far end of the field with our wing tip. This swung us round and down to hit the ground, tearing off the undercarriage. As two gallant and red-faced British airmen scrambled, shaken but unhurt from the bent aircraft, the children trooped back to their classes, without a second glance.

The plane wasn't badly damaged and, after repair, it was flown back home to Montrose.

Brian had been lucky, compared to many of the young pilots. He didn't want to worry his sister, so he told her that there was no danger as the planes were too low and too slow for anyone to be hurt. However, there is no doubt he would have been aware that, on average, there was a crash every day and a funeral every week – as the military graves in the nearby Sleepy Hollow cemetery attest.

Events were moving on, with new aircraft arriving almost daily at Montrose. Throughout these few weeks since his transfer to the RFC, Brian and his fellow officers heard news of the war across the Channel and their compatriots in the air and on the ground fighting the German invaders. He knew it would not be long now before he joined the fray.

Chapter Two

1915–1916

Bombing Blind

It was early December 1915 when Brian and his fellow pilots received their orders. After a brief home leave in Hertford and some emotional farewells from his mother and his sister, Brian returned to Montrose, where he joined his friends in the RFC's recently formed No. 25 Squadron.

The following morning they took to the air and flew south to Thetford Aerodrome in Norfolk to refuel. They must have looked a motley crew, flying an assortment of open-cockpit aircraft, including S7 Longhorns, Curtisses, Maurice Farmans, Martinsyde Elephants, Caudrons and Avros. These young pilots must have had misgivings about taking such primitive planes to war, but on arrival at Thetford the rumour was that most of them would soon be replaced by the brand-new two-seater FE2b, the Bristol Scout and the BE2c fighter. This latter was the one Brian coveted, but so did most of his pals. They knew these more operational aircraft would be a great improvement in every way.

Brian and some of his colleagues were excited at the prospect of active service, yet all felt some degree of apprehension as they bedded down that night in their hut. The tension rose when they woke the following morning and prepared to leave. But their spirits were greatly boosted when the station commander brought them together to wish them all Godspeed … then added that their new aircraft would follow them in a few days' time. No doubt their cheers resounded behind them as they climbed into their old machines for the most important journey of their lives, ready to serve their country.

So it was that on 2 December 1915, with less than twelve hours' solo flying practice behind him, 18-year-old Brian Baker took off to join the war with a Smith & Wesson pistol on his lap and an observer behind him with a rifle across his knees, ready to watch out for enemy snipers as they crossed the Channel.

As Brian's pilot's flying log book shows, in his spidery writing but with a firm hand he recorded the date, the 'light wind' and his destination: 'French aerodrome'. When they arrived and started their descent, the view was so different from their familiar patchwork of green fields at Montrose airfield. Here they saw a barren-looking plain surrounding what looked like an extensive hard surface, about a kilometre square. There was very little greenery in sight, other than some mostly leafless winter woodland. As Brian brought his plane closer to the ground, he spotted a little chateau on a hill at one end of the aerodrome, between it and a small town. This chateau was to be the squadron's headquarters.

Saint-Omer Aerodrome was situated about 40 kilometres south-east of Calais. Both the town and the aerodrome took their name from the seventeenth-century Saint Omer and his monks, who built a chapel in the marshes nearby. Most notably, 200 years later, the aerodrome was the site of a Napoleonic army camp, where Napoleon himself planned his invasion of England, which was aborted soon after.

Once all the aircraft were safely on the ground, mechanics took care of them while the men turned their minds to finding billets. If anything, this now seemed to be even more challenging than their journey. There was very little accommodation onsite, and that was fully occupied, so Brian and his companions had to be billeted in local people's homes in the town, which was often not popular with their hosts, who found some of the airmen's habits 'distasteful'. An HQ officer warned the newly arrived airmen about this, telling them of a recent incident. Two elderly ladies had come to the chateau with a 'delicate' complaint about their RFC lodger who dared to wash his socks in the kitchen sink and hang them there to dry. This caused some merriment among the new arrivals, but it made the point well!

The following morning, as the pilots reassembled at one end of the aerodrome expecting to receive their orders, they heard a distant hum, which was growing louder. They recognised it as the sound of approaching aircraft. Would this be the hum of enemy planes? A few of the mechanics turned, ready to run, but there was nowhere to hide. Just then, a couple of sharp-eyed young airmen whooped and pointed at the leading aircraft. By now, everyone had noticed the RFC markings. What a welcome sight it was – a flight of brand-new aircraft circling the aerodrome. As they dropped height and came in to land, the men flocked round, eager to inspect their new steeds. As soon as the planes landed, the pilots gathered round to inspect them, sharing their enthusiasm as they pointed out to each other the new features they had heard about.

Once the excitement had calmed sufficiently, their commander read out the list of how the aircraft would be allocated. Brian must have cheered when he heard his name and the model allotted to him – the BE2c. He and his fellow pilots spent the next couple days familiarising themselves with the controls and the way their new machines responded, proud at the prospect of serving their country in aircraft befitting the task. Brian wasted no time. In the first days after his arrival, his log book lists a number of short flights that he took around the aerodrome to start with and then across the local countryside, learning the place names and their geographical features. Of course, Radar hadn't yet been invented, so inevitably in those early days pilots would often lose their bearings over the unfamiliar terrain. As Brian later wrote:

The answer was to find a railway line and follow it until you came to the station then dip down to read its name. Alternatively, if the weather was bad or visibility was very poor we landed as soon as possible, then looked for a farm-house and a bed or a barn for the night.

In the first few weeks, the pilots had teething problems with some of their aircraft:

If our engines cut out, and they frequently did, we continued through the air and glided down quietly in a field. Engines were stopping all over the place and extraordinary things happened. I remember one time when I was standing on the tarmac, watching a Curtiss ticking over, waiting to take off when, all of a sudden, the whole engine fell out on the ground. The C.O. came over to see what had happened. 'Good gracious,' he said. 'Look at that mess! Where's the Technical Warrant Officer?' The T.W.O. came along in a jiffy and, after a quick glance, he said, 'Well sir, I can see exactly what's happened. Some damned fool hasn't drilled the bolt-holes through the middle. He's drilled them at the side. And it's just sheared off, letting the whole engine fall out.

There wasn't a court of enquiry or anything. They just swept up the engine and took it to the workshop.

Only a few days before, Brian's entries in his log book had detailed a flight in a Curtiss at Montrose in which he rose to a height of around 2,000 feet, but now in his BE2c he was taking 'testing engine' flights, high in the sky at up to 9,000 feet over Amiens, Cambrai, Albert and Beauval. It was all good preparation for his first big mission – to bomb Cambrai railway station in order to prevent German soldiers from being sent to strengthen their presence on the Western Front.

None of Brian's cohort had yet done any bomber training, so he had a quick run through with a more experienced officer on how it all worked and what he had to do, but he knew it would be a very challenging task with such rudimentary instructions and unfamiliar equipment. With a mixture of apprehension and eagerness to contribute to the war, he took off early the next morning in a clear sky and headed for Cambrai. In his own words, written some years later for a newspaper, Brian described this experience:

At that time, I had to leave my observer behind as we couldn't carry the weight of the bombs as well as an observer. It had to be one or the other. So my twelfth hour of solo flight was spent over

Cambrai Station with two 12-pound bombs strapped beneath my seat. I hoped I dropped them onto the station. It was difficult to be certain, because we had no bomb sights. You simply looked over the side, lined up your target, pulled a little toggle and the bombs fell off.

Brian and his fellow pilots and observers were now on regular duties, mainly on reconnaissance down the German lines. It was dangerous work, not least because the German aircraft were so much more versatile than the British planes. As Brian himself later wrote: 'At that time we were flying B.E.2c aircraft, which were no match for the nippy German Fokkers, which could fly faster and make a tighter turn. Their all-round advantage meant we had to be better pilots to say alive.'

What these young pilots did not know at this early stage of the war was that the average survival rate of new RFC recruits going to the Western Front was just six weeks ... and not much more for all the other airmen.

Brian's first letter from France to his sister gave no hint of his concerns:

24 January 1916

My dear Win,

Thank you very much for your letter and Mother also for her card. Thank Dad for the smokes. I have quite a tobacconists shop now.

We were inspected yesterday by Sir Douglas Haig and crowds of brass hats. They arrived in a train of motor cars, which followed them about, wherever they went.

There is a French motor-school next to us and they went there first. We soon knew when he had arrived there because there were two of the most terrifying crashes I've ever heard. The whole aerodrome shook all over.

It was a very foggy day, so we did no flying for them to see, but we had to be working in the sheds, or looking as if we were.

I have got a splendid room and bed to sleep on, but to get a bath is an awful trouble. Will you send that bath affair out here sometime? Then I shall manage all right.

Those machines that you saw going over Hertford were the new ones that 20 25 were taking from Farnborough to Hertford, I learned from Ridd yesterday. He told me they were coming out here about Feb 10th.

We are going to be formed into a new squadron, spotting for artillery, as far as I can make out. It is a far safer job than doing long reconnaissances about 50 miles into Germany, because we'll only have to go just over the lines, so that we can always get back if our engine konks.

Love to all,

Yours affectionately

Brian

In another article, Brian wrote in more detail about his first weeks of the war, explaining what some of his work entailed:

When I went to France, I knew I would be going into combat. Having received no special training in armaments, bomb-dropping or aerial warfare to prepare us, it was just a case of using our common sense and initiative to stay out of trouble and, if we survived long enough, developing our own techniques.

R.F.C. operations in those early days were very haphazard. The problem was that, as flying was very much in its infancy, nobody had any real idea of how to use aircraft to their best advantage. Powered flight by man was absolutely revolutionary and there was no yardstick against which to measure it.

When the war began, there were only four squadrons in the whole of the Royal Flying Corps, although it quickly went up to six, with a total of something like 50 serviceable aircraft. Instructions for reconnaissances came down from Wing Headquarters, after they had received them from the army commanders. At that time,

Wing H.Q. was part of the army and did what they ordered. We were regarded solely as the eyes of the army.

Initially, we went up on a variety of jobs, including reconnaissance, bombing and aerial photography.

It was a great day when we got a Lewis gun with mounts fitted to our aircraft. The observer carried it on his knees. For firing ahead he mounted it on a bar across the two front struts. If we wanted to have a go at a plane on our tail, he would put the gun on a tripod on the top of the fuselage behind him.

As the observer sat in front of the pilot this meant the man at the controls was virtually staring down the barrel. Unfortunately it was about a foot above his head. Even so, the concussion of the bullets firing felt rather like somebody tapping you on the head with a hammer.

The main problem about firing forward was not to shoot off your own prop. Firing at something below wasn't too difficult since the observer could aim between the propeller and the front [of the] plane.

In order to stop him shooting away the prop as he moved the gun up and down to follow his target, we had two pieces of ordinary linen tape nailed from top plane to bottom plane. As long as he didn't pass out, we were all right.

To begin with, the Germans had a machine gun firing through the propeller, which had a deflector plate, in case a bullet struck it. They did however have an interrupter gear, so that the majority of bullets passed between the blades, but the plate was there as a precaution.

Our fortunes improved with the arrival of the new FE2b planes from Montrose. These were two-seater Pushers. The observer sat in the front with a gun, while the pilot sat behind him. These planes were a definite improvement, while they lasted. When attacked, their method was to go round and round, one behind the other so they always had a gun covering the tail.

Then we got the DH.2, a single-seater fighter with a gun mounted in front of the pilot. It was a lovely little thing, very strong and could dive the wings off a Fokker.

If you got on his tail and he tried to escape by diving, the DH.2 was stronger. If you hadn't shot him down first, the Fokker's wings would break off under the strain.

The DH.2 was really the first scout or fighter, as it became called later. It had a top speed of 100 m.p.h. and was the forerunner of the Bristol Fighter, SE5 and the Camel.

Brian also did a lot of aerial photography, or rather he took up an aerial photographer in a Bristol Fighter at 20,000 feet. The reason for this height was to take photos of whole areas of enemy positions, warfare and landscape features for use by the British forces. It was a more detailed method of reconnaissance and an essential aid to the Allied planners. However, at that height, there were inevitable problems to surmount, as Brian described what happened on one of his sorties:

We had no oxygen and it was pretty chilly, sitting in an open cockpit despite fur-lined coats, helmets and flying boots. Breathing wasn't quite such a problem for the pilot, because he didn't have to move around, using up energy. But the poor photographer was up and down all the time, removing exposed plates and reloading the camera.

It was quite an effort for him, handling the heavy equipment.

As a result, the pilot had to keep a close eye on the photographer. It wasn't unusual to glance round and find him crumpled in an unconscious heap.

Being a modest man, not given to boasting, Brian chose not to exemplify this scenario in his newspaper account. However, the very same situation did happen to a photographer in his plane. He couldn't rouse the photographer with shouting or nudging. Several seconds had passed

– perhaps too many – so there was only one thing Brian could do. Unable to leave the controls, he deliberately put his plane into a steep dive, as fast as he could, down to 10,000 feet, and then stabilised long enough to look round again. For a moment, the hapless photographer was still in a motionless heap … then a jerk and he suddenly came back to life, stunned and disorientated … but alive.

Brian was elated as he headed the aircraft back to base. His quick decision to drop the plane so dramatically that the atmospheric pressure shocked his photographer's system had saved the man's life.

Brian's pilot's log book shows that he flew every day and often several times a day throughout January 1916, bombing targets, carrying out reconnaissance within enemy lines and fighting Fokkers in the sky. On one of these occasions, Brian saw a British fighter being chased by two Huns, so he went to his aid. However, before he could join the fray, his compatriot took a direct hit, setting his plane on fire. Brian was horrified to see what the flames did to what he could see of the pilot's face as his plane went into a spin and spiralled down to crash on the ground.

Brian knew this might have ended differently if the War Office hadn't downright refused in those early days to fund the provision of parachutes. Apparently, the officials believed they would not be of much use. The rumour among the pilots was that the 'boffins' thought that if the men had parachutes, they would jump out, though of course that could have been suicide at those heights.

Brian's next letter home lacked his usual ebullience:

4th February 1916

My dear Win,
Many thanks for your letter and please thank mother for hers. I have been in bed for the last two days with something that gives me an angry stomach. The doctor is coming this afternoon. I expect I shall be pushed off to hospital.

It has been much cooler just lately here. I saw Tony again a few days ago, just as he was off to the trenches. I don't envy him.

I must stop now as it makes my arm ache to write.

Yours affectionately,

Brian

No matter how worried they were, or how many letters they wrote to ask what the doctor had said, the family now had to wait sixteen days before their next letter from Brian. It finally arrived on 20 February from an unexpected address: The Michelham Convalescent Home for British Officers at Cimiez, near Nice in Southern France. From here, Brian wrote a very long letter, waxing lyrical to Win about the weather, the semi-tropical flowers and fruits, the exhibits in both the Monaco Museum and its Aquarium. He goes on to describe the food and that he was sleeping in the annexe, which had previously been occupied by Queen Victoria – maybe even in the same bed!

20 February 1916
The Michelham Convalescent Home for British Officers, Cimiez, Nice

My dear Win,

I hope you received the postcards that I sent yesterday. I've got some more in this place which I will send soon. It's simply glorious down here, just like our August, but directly the sun goes down it gets very cold.

The sea is an amazing colour … palm trees and orange groves all over the place.

I sleep in the annexe of the hotel, which is the house that Queen Victoria used to live in when she stayed down here. …

The place is run by Lord Michelham and they do us awfully well. The food is simply A1 and drinks at meal-times are free. They take us out for motor drives all around the country and give us tea on top of it all.

Roses and all sorts of brightly coloured flowers are blooming all over everywhere.

There are 130 officers here all told and they keep you here between three weeks and a month, then foist us back again.

The most amusing thing is that hardly any of the people here look as if there is or was anything wrong with them.

Brian writes some more about his surroundings and views, then finally closes without mentioning a word about what is wrong with him and how he is progressing.

… Nothing more to say just now.
Love to all,
Yours affectionately,
Brian

Nine days later, on 29 February, Brian wrote another very long letter, describing outings to Monaco and a hair-raising motor trip, being driven up tortuous mountain roads with hairpin bends, too close to the precipice. 'One fellow with nerves went and he said he would never go again.'

Brian has taken an interest in what foreign armies are joining the war and training nearby, especially the Indian Cavalry Corps, who 'spend most of their time drilling behind the lines, charging at hares and partridges with lances'. His letter ends with the weather, the wildlife and almost everything else, but still nothing about himself, and what he is convalescing from.

Five days later, Brian had travelled north and crossed the Channel to take up a bed in a military hospital in London.

22nd March 1916
1st London General Hospital, Camberwell

We took a most extraordinary course across the sea, changing direction several times. We arrived at St Margaret's Bay, where the old four-masted ship is still there on the rocks. I was pleased to see Blighty again, I can tell you.

The Doc has just been. He didn't say much. I have got to stay in bed today – Heaven only knows why? …

I shall be mighty pleased to be sent off hospital and come on leave as I am fed up with all this messing about.

Yours affectionately,

Brian

The family photos, all labelled by Win, show that Brian finally had his time at home on a long leave with his family. However, in April 1916, Brian's father sadly died at the age of 61, which must have been a shock for them all, but at least Brian was there to mourn with his mother and sister.

Looking closely at one or two of the photos we can see how thin Brian was when he first arrived home, but as spring turned to summer, he soon picked up his strength and health. By early October, he had been posted to Spittlegate Camp, Grantham, where he was so bored that he stripped down his motorised bike 'into 100 pieces' and rebuilt it, waiting to get back into active service.

Finally, Brian was passed fit by the Medical Board. Now all he was waiting for were his orders to rejoin 48 Squadron and get back into action against the Huns. As he wrote to his sister: 'I can't wait, with terrific excitement for work.'

Surely, it wouldn't be long now.

Chapter Three

1916–1917

Bait for the Red Baron

Instead of going straight back to the Western Front as he had expected, in December 1916 Brian was sent to Catterick and also Castle Bromwich as a flying instructor. He enjoyed teaching others, but he was impatient to get back into action … and it wasn't long before his wish came true, when he met an old friend, as he recounted in a letter to Win:

> Who should land here but Col. Holt? He was wonderfully pleased to come and see me, he said, and asked me if I would like to come into Home Defence. I thought it was too good a chance to miss, so I told him I would.

On 9 January 1917, Brian was promoted to captain. Five days later, he flew himself down to Filton Aerodrome, near Bristol, to join Home Defence, with orders to report to No. 35 Squadron. His first problem was that there was no 35 Squadron yet. His second problem was that he was the only officer on the base, with forty men eagerly awaiting their orders. Apparently, the CO was due to arrive later in the day, so Brian gave the men a few hours off at the aerodrome while he gave himself a day's leave and headed for the nearby Park Hotel, where Sir Fred, a family friend, was staying.

Brian was now given command of the newly formed squadron, guarding the country's southern shores. He and his men patrolled the coasts, the Channel and beyond, into Normandy, chasing off enemy aircraft and the Huns' treacherous airships that dared to trespass into

England's airspace, carrying their bombs. This was more than a full-time job for all the pilots, often involving dangerous dogfights in the skies before they had to fly all the way back to base, where they would fill in their log books, have a quick meal, then bed and out again by dawn. No wonder he complained in his letter to Winifred of 14 March: 'We start flying at 6.30 in the morning and go on till about 6.45 at night, so you can imagine that we get a bit sick of it.'

April 1917 would become known as 'Bloody April', for good reason. A fierce battle was raging near the French town of Arras, involving the majority of army and air force personnel from both sides. As well as the essential reconnaissance flights, the British airmen spent their days bombing the German trenches and artillery positions, covering a vast area of land and the inevitable fighting in the air overhead.

Despite the Allied airmen's strong spirits and skills, their aircraft were not up to the job, so the Germans in their more advanced fighting machines had the upper hand. Brian's 48 Squadron and twenty-four other British squadrons, involving 365 aircraft, were deployed during this long and bloody combat. Many of the British planes were outdated and outclassed, with the result that, over this year, the RFC lost 245 aircraft, 211 airmen killed or missing, and 108 captured as prisoners of war. Fortunately, with his trusty Bristol Fighter and considerable expertise, Brian survived the aerial warfare of Arras unscathed, at least physically.

Despite the Allied losses, the Battle of Arras became a turning point in the war. Within two months, Britain's technological advances had produced a new generation of fighter aircraft, including the SE Sopwith Camel and the SPAD S.XIII, taking supremacy over the German Jastas. As a result, British losses fell and German losses rose.

On 19 April 1917, after a short home leave, Brian returned, this time by sea, to the Arras battlefields and a new base.

29th April 1917

My dear Win,
I had quite a good crossing and met old man Sharpe on the boat.
We went over with hospital ships and transports and 3 destroyers,
as well as an airship.

There was an awful good fellow who used to be at Mondemont
who came over with me and we had a great night out in Boulogne.

There are still the same old cheery crowd in the squadron and
a lot who were at Filton.

The huts here are quite good ones, but are pretty cold in the
mornings. The building is simply terrific in the morning and
evening and there is a dance every evening.

We had a good scrap this afternoon – three of us and five Huns.
We sent one of them down through the clouds in an awful hurry,
but we couldn't see if he crashed.

Love to all.

Yours Affectionately,
Brian

In Brian's pilot's log book for that day, he described this aerial encounter
in his Bristol Fighter as 'Fight with 5 Huns'.

From May 1917 onwards, Brian's pilot's log book shows that he was
flying every day, several times a day. In a newspaper article he wrote
many years later, Brian looked back at this middle period of the First
World War, when he described the ways aerial battles developed, along
with the manoeuvrability of the aircraft and the skills of the pilots
who flew them:

In the First World War, aerial combat was conducted in a very
gentlemanly fashion. It was one against one, round and round you
went. First you got a pot at him, then he got a pot at you. Neither
side could carry much ammunition and one was bound to run out
before long. Then it was a case of taking desperate evasive action,

hoping to survive until your opponent also used up all his bullets. Once that happened, you just gave each other a wave and broke off. He roared away in one direction and you roared away in the other. Perhaps you'd meet again the next day and start all over again.

In these dog-fights, we were at a bit of a disadvantage. The wind nearly always blew from the west and, as we kept going round and round, it gradually drifted us east further into enemy territory. So we had much further to fly to get home and the petrol situation could be critical.

I got caught up in quite a few good-going scraps and luckily I was never shot down, but my aircraft was often full of bullet-holes when I got back to base. The closest call I had was when a bullet from a rear gunner zipped cross the top of my head, digging a groove in my leather flying helmet. I felt as if someone had battered me with a brick and in fact, I blacked out. Fortunately, I must have come round within seconds because the plane was still the right way up and I don't think my observer even realised what had happened.

Some of the dog-fights were spectacular affairs, involving twenty to thirty aircraft, criss-crossing the sky at varying heights, with their own private little feuds going on for fifteen minutes or so. Then within seconds it was all over – the sky cleared and the survivors scuttled off home.

Brian also wrote about the standout characters on both sides and their modus operandi. In particular, he admired the most successful air aces of his time, their courage and gallantry as much as their victories:

Our fortunes improved with the arrival of the new F.E.2b planes – two-seater pushers. The observer sat in the front with a gun, the pilot behind him. They were a definite improvement, while they lasted. When attacked, their method was to go round and round, one behind the other, so they always had a gun covering their tail.

Then we got the DH.2, a single-seater Pusher scout. It was a fighter with a gun mounted in front of the pilot and the propeller behind. It was mainly ash wood with spruce for the wings. If you got on a Fokker's tail and he tried to escape by diving, but the DH.2 was stronger. If you hadn't shot him down first, the Fokker's wings would break off under the strain.

The DH.2 was really the first scout, or fighter as it became known as later. It had a top speed of 100mph and was the forerunner of the Bristol Fighter S.E.5 and the Camel.

The S.E.5 was the plane in which the famous 'Aces', Ball, Bishop, McCudden and Mannock, did such destruction against the enemy – not because they were exceptionally brilliant pilots. In fact, there were better pilots in the squadron, but the Aces were extremely good shots and their methods marked them out.

Ball was a very gallant young man, but he was impetuous and rather rushed into trouble. He was bound to go sooner or later … and he did, when he took too great a risk over enemy territory. Much later we found out the Huns respected his prowess so much that they buried him with full military honours. Gallantry was a shared virtue in those days.

By contrast, McCudden was a most calculating type and he spent more time in the air than anybody else. He was a fine mechanic, who looked after his own engine and guns. He even balanced his own propeller, so that he could get that little bit extra out of it for height. He spent hour after hour in the air, just sitting there, waiting to get a chance at the enemy. The chances came and, being an excellent shot, he took them. That's how he amassed his 70+ victories.

Bishop was a Canadian, a former Mountie who found flying more exciting, whereas Mannock was known to be a sensitive type, but soon found his form once he'd shot down an enemy balloon.

These four 'aces' were joined by a new ace, Brian. All of them were officers, very close in age and keen to collaborate in order to gain more victories. In particular, they all shared the same goal – to rid the sky of the flamboyant Prussian aristocrat Baron Manfred von Richthofen, also known as the 'Red Baron'. He was the most prolific enemy, in the best aircraft, and had gathered a group of his most successful companions to join him, flying in formations of four or six fighter planes, all of which were painted red, to identify himself and his team and, even more importantly, to put the fear of God into any Allied pilots who should dare to fly up when the Red Baron and his 'Richthofen Flying Circus' were in the sky.

In his newspaper article, Brian explained his key role:

On a number of occasions, I became the bait for Bishop to lure the Red Baron himself.

That meant me going out in the early morning in my Bristol Fighter and trying to attract the famous Richthofen's Circus, while Bishop sat high above me, in the sun, waiting and watching.

When they turned up to have a go at me, he pounced.

I must admit, I always prayed he would pounce quickly enough, because the Red Circus were pretty hot.

My main aim was to make sure the enemy aircraft were below Bishop. That was the basic rule of our dog-fights: Get above your adversary and you had him.

There was a different technique for dealing with the German two-seater reconnaissance aircraft. You approached them from underneath, so that the rear gunner couldn't get a shot at you. You could creep up until you were close enough to put him down without any fear of return fire.

By this stage of the war, Richthofen's name was almost as well known and feared across Britain as he was known and admired in his own homeland. However, as Brian wrote:

We soon discovered Richthofen didn't quite live up to his glamorous reputation. He had a little trick of shooting over our lines at very low altitude, early in the morning and pouncing on a lone reconnaissance aircraft, which he would shoot down within a second. Then he'd high-tail it back before we could go up to interrupt him. He notched up a lot of his victories in this fashion, which we regarded as 'not quite cricket'.

Finally, in July 1917, Richthofen's unsportsmanlike behaviour went into an ignominious decline due to his sustaining severe head injuries in an attack by one of Brian's fellow pilots, Captain Donald Cunnell. In between surgeries, the Red Baron re-joined his circus and downed more Allied planes, but not for long. He died soon after from the delayed effects of his injuries. He was 27.

16th May 1917

My Dear Win,
The weather's got a bit dud at last, so we're having a rest, which makes everybody extraordinarily cheerful. I am Squadron Commander for the day and, as the Major is away, I am not allowed to fly, so I stay on the ground, look after the place and send the other poor devils into the air.

We had a little ammunition on fire the other day. You've never heard such a row in your life. Bombs are like crackers compared to it.

A few days ago we had a turn firing into the trenches. It was an awful good show and put the fear of God into the Bosches.

Brian's next letter demonstrates how, although he enthusiastically relates the small day-to-day happenings on his base, he refrains from telling his family the more dangerous aspects of his war and the many close escapes he has had in the air, earning much praise from his superiors.

4th July 1917

My dear Mother,

Just a line to tell you that I have got a Military Cross. What for, goodness knows, but I got a telegram today to say I had won one.

We are in the midst of packing up for another move and the deuce of a job it is too. My address will be just 48 Squadron, RFC, BEF. Leave out 13th Wing and then it will be all right. I am going to move as many flowers as possible in the lorries, but as I know the people in the squadron that are going to take over from us I shall leave some of them for their benefit.

We are going to a much nicer place than this, so we are not grumbling.

There's nothing more to say and I am rather busy.

Love to all,

Yours affectionately,

Brian

The citation for Brian's Military Cross tells a braver tale:

2nd Lt. (T./Capt.) Brian Edmund Baker, Rif. Bde., and R.F.C. For conspicuous gallantry and devotion to duty. He led his patrol with great skill against a hostile formation, which he attacked, accounting for five enemy machines out of six. Later, he drove a hostile machine down in flames, and attacked and destroyed another one by diving 7,000 feet on to it and firing at such close range as to nearly collide with it. His gallantry has been at all times of great value to his squadron.

This award was announced in both the local and national newspapers but Brian himself didn't see the certificate, let alone the wording, until he came back from an advance party visit to the squadron's new base. As he wrote to his sister, Win: 'We are going to a grand place, but it's tents, not huts.'

Brian later wrote in an article:

> One July evening in 1917, along with other members of 48
> Squadron, I was on photo-reconnaissance patrol off Ostend, we
> sighted a lone German Gotha bomber coming in from the sea.
> We deduced rightly, as it turned out, that he was returning from
> a daylight bombing raid on London. We closed on him as fast
> as possible, fired a few bursts and by the time we turned to have
> another go, he was already down in the sea. We spotted his gun
> quite clearly on the beach, the aircraft's tail sticking out of the
> water. That 'kill' gave me a lot of satisfaction. We later found out
> that about 46 enemy aircraft had bombarded London that day,
> killing 94 people, 26 of them children.

In a letter to his mother, Brian explained the actions that led to his MC
and how they were mostly shared victories. He finished with: 'This
aviator was ranging Hun guns on us and had been doing an awful lot
of carnage for some time, so I was rather lucky. For goodness sake,
don't let them make such a song about it.'

On 26 July, Brian wrote home to tell Win and his mother that
General Rawlinson had come that morning to invest Brian with
his MC, in front of the whole squadron, some of whom had also
gained decorations:

> It was a wonderful show and amused me no end. Nobody made
> much of a howler, but I shook hands with the old boy twice instead
> of once, and another gent saluted with the wrong hand, which
> aroused a little suppressed mirth. He didn't give me a medal, so I
> suppose that will come in parcel post.
>
> Now that it has been put in the paper, it might be of interest
> for you to know that I shot a Gotha down in the sea. I told you
> that we had a colossal morning on Sunday. Well that's what it
> was. He went well in, with his tail up in the air.

The Hun communiqué says one of our machines 'fell into the sea, going to an airborne cause'. I suppose I was the 'airborne cause'.

We had lovely bathing in the sea this week. The jelly-fish have all pushed off, so it is ripping now and beautifully warm. All the time we were bathing, we could see the aircraft over the line being crashed like anything. It seems so funny to be engaged in a peaceful swim, only a few miles away from the front line.

Love to all,
Yours affect.
Brian

After receiving his MC, marking his first successes as a flying ace, Brian was always keen to fly, whether in a group or alone, hungry for more victories. His pals told the story of one memorable early morning when Brian emerged from his tent in his pyjamas and announced, full of enthusiasm: 'I've got to go up.' He proceeded to pull on his flying kit over his pyjamas, ran to his plane and took off.

Twenty minutes later, Brian returned, exclaiming, 'I got it!'

Indeed, the plane he claimed to have downed was spotted four hours later by two of his friends just 50 yards from the shore, where its wreckage was still sticking out of the water between Nieuport and Ostend. Having previously accounted for two planes at once, this was Brian's third victory. To add to his celebratory mood, news came through in July 1917 that his most formidable enemy had been shot down and badly injured, so he would be out of the sky for a while, at least.

With his favourite plane yet, the Bristol Fighter, Brian notched up several more aerial victories.

'All pilots know what it is to be frightened,' Brian had written earlier in the war, when the German aircraft were vastly superior to Allied planes, but now, at last, this tide was turning in the Allies' favour and their newfound confidence in their machines reduced the airmen's tension.

Chapter Four

1917–1918

Beating the Baron

'Number 48 was truly a remarkable squadron. It had quickly acquired a reputation commensurate with the first Bristol Fighter Squadron ... the three flight commanders were all distinguished in their own way ... Brian Baker, then just nineteen and dubbed by his contemporaries, for reasons now obscure 'the corporal-major', had a brilliant career ahead of him.'

(Ralph Barker, *The Royal Flying Corps in World War I*)

B rian and his companions worked hard and often long hours in the sky, but they relaxed whenever and however they could.

We were stationed at Dunkirk and when we weren't flying we had quite a riotous time visiting other aerodromes or the Chateau Rouge in town. Or sometimes we went down to the beach.

The Germans had a monstrous gun called Mournful Mary. From it they could fire shells into the middle of Dunkirk, more than twenty miles away. One of our entertainments was to go out on the airstrip at night and sit on some sandbags, watching for the flash when Mournful Mary was fired. From that moment it took exactly five seconds for the sound of the gun to reach us, and about 25 seconds after that the shell passed overhead en route for Dunkirk.

One early morning, when Brian was flying with a group of his pilots, he spotted in the distance a formation of enemy planes flying towards them. As they approached, the sun glinted on one aircraft's red wings, which could mean only one thing: Baron Manfred von Richthofen and his Flying Circus were on the prowl.

As they drew closer, the red plane took the lead. Could it be the Baron himself? Yes, there was no doubt of that.

As flight commander, Brian instinctively did the same. He knew it could be a foolhardy move, but he had tussled with the Baron before … he couldn't resist the challenge.

As the followers on both sides held back, Brian took the initiative and flew up above his target, from where he could see the Baron's personal crest. Then Brian dived and, in that split second, the Baron uncharacteristically hesitated. Next, he turned to one side and dropped his altitude, perhaps to give himself time for a counterattack. But it was already too late for that – and too close to a French hamlet beneath them – so the only thing Richthofen could do now was to flee.

Brian's instinct was to dispatch the Baron for good, but that would have caused a crash and the loss of lives below. Although Brian was disappointed that he hadn't shot down the arrogant Baron for good, he was glad that he had at least chased him out of the sky and given him a fright to remember.

There were, of course, days when Brian was the one under overwhelming attack, as he related in one of his newspaper articles:

On a clear day, I went up from our base at Dunkirk to take some photographs of the Belgian coast near Ostend. My photographer and I were so busy concentrating on what we were doing, we forgot to keep a weather-eye open for enemy aircraft.

Suddenly we were jumped by three German Albatross fighters and things became hectic. Throwing my aircraft around in the sky, I was having great difficulty keeping us out of their gun-sights. The outlook was bleak.

Out of the blue, a small Belgian fighter hurled itself into the melee. The Germans seemed to think he was the advance guard of a Belgian flight. Immediately they broke off and scattered, leaving us shaken but unhurt.

The courageous Belgian pilot escorted us back over our lines and landed at our aerodrome. He climbed out, came over and asked if I was alright.

'Yes indeed,' I told him. 'Thanks to you.'

We repaired to the mess for a drink and I discovered his name was Willi [Willy] Coppins.

For that brave intervention, he was awarded the Military Cross (MC). He later became Belgian attaché to London. I looked him up after the war and we had lunch together, discussing those days.

After several more occasions when Brian shot down enemy planes, scoring sixteen victories in all that year, his commanding officer handed him a French telegram. He gazed at the official-looking crest on the front, and then opened it to find that it was from the President of France, awarding Brian France's top military honour, the *Croix de Guerre* with palm.

'I can't think why I got it,' he wrote to his mother a few days later. 'The Colonel presented it to me, so at least I didn't have to be kissed.'

As the 1917 autumn weather grew colder and foggier, visibility dropped, so there were many days when the pilots couldn't fly. Brian visited other squadrons and met old friends, including one fellow old boy from his school, Haileybury College in Hertford. They had much to catch up on and, as he was an army man, Brian took him up for brief a spin in his plane. On other days, Brian and his pals put on some sports matches. Brian had always been a keen sportsman, especially an outstanding cricketer, but at this time of year, it was his second favourite sport – rugger. They played other squadrons and the score line was almost always in double figures for Brian's team and nought for their opponents.

Extracts from Brian's letters to his sister give a taste of his daily life between sorties:

24th September 1917

My Dear Win,

Sorry I haven't written before but I've been very busy and have been out a great deal … We had a grand bathe this afternoon and a colossal air-raid this evening. It went off all night …

It is terrible cold at night. Another fine fat Hun flying over – great excitement.

9th October 1917

My dear Win,

I haven't heard from you for some days now, so I suppose you have been Zepped again. They brought off a great show out here. The British did in all 69 of them. The Huns all got lost and finished off somewhere in the South of France.

The weather has cleared up a bit but I don't think it will last.

27th October 1917

I have four horses attached to the platoon and have been learning to ride. In my spare time I am getting quite expert at the art. Every morning we can get off, we have terrific gallops across the sands before breakfast. I expect to fall off any day!

Sunday flying.

Wondrous cold.

The enclosed is from a magazine, which might amuse you. I shouldn't show it to Mrs. Webb, or you may have to carry her off.

Everybody is very excited over the push. It was a good show – it took the Bosch by surprise and knocked them all over the place before they could look round.

29th November 1917

A Frenchman has just written off a machine in a ditch beside the aerodrome, but he didn't seem to hurt himself. The machine is all smashed to pieces. This week we've put on concerts, dinners and plenty of bubbles water.

26th December
We had an excellent evening last night. A wonderful feed and plenty of it. I hope you had a good time. The snow is something wonderful – 6 inches deep and drifts of about 3 feet. I've just got back from viewing a crash, but couldn't get near it.

We are now billeted in a fine chateau – an enormous place with velvet on the walls and fine fireplaces. It belongs to a baroness, but she has locked up all the crockery, which is rather a nuisance. We all have magnificent spring beds, which is a great asset.

This relatively calm period in late 1917 was deceptive. It was the proverbial calm before the storm.

Brian had expected to go on leave over the Christmas or New Year period, but that kept being postponed and it would not be very long before he found out why. But first came a surprise that he shared with his sister on 13 January 1918:

My dear Win,
Just a line to tell you that I have just got a DSO (Distinguished Service Order) for Heaven only knows. It was a great surprise.

We are flying again now, after a bit of a break, but the schedule has changed and it really does not sound good.

I have got the General coming tomorrow night and we have raised a band from a Cavalry Corps close by.

We ought to have a good evening.

Give my love to all.

Yrs. affec.

Brian

Brian's DSO citation was signed by Sir Henry Rawlinson, Commander of the Fourth Army, and read:

T./2nd Lt. (T./Capt.) Brian Edmund Baker, MC, Gen. List and RFC.

For conspicuous gallantry and devotion to duty. Whilst on patrol he engaged nine Albatross scouts, five of these being driven down, two of which he accounted for. On another occasion, whilst leading his flight on an offensive patrol, he dived alone on a formation of six enemy scouts, driving one down out of control. During the course of his patrol work he has brought down ten enemy machines and his work on all occasions has been magnificent. He is a dashing patrol leader, and inspires all with the greatest keenness.

The Colonel arrived a few days later, so another dinner and band were enjoyed by all.

Judging by an extract from the letter Brian wrote to Win on 5 February, his family were not unscathed by the war:

Thanks for your letter I received tonight about the air-raids. What a terrific show it must have been. We have had one or two shows lately, but they were not very exciting. Not a match on our old ones. But there is plenty of time for them to do more.

There follows a three-month gap before Brian's next letter.

Rumours had been rife for some that the Germans were hatching a major plan to mount a spring offensive against the Allies, so all RFC personnel on the Western Front had their leave cancelled.

The Kaiserschlacht, its official title, or its German code name Operation Michael, was already building up. Intelligence came in that seventy-two enemy divisions were positioning themselves, ready to attack the Allied forces in three waves. The Allied infantries of Britain and France, with valuable support from Australia and America, mobilised to the Somme battlefields. Meanwhile, the Allied air forces equipped themselves and stood at the ready.

In an article Brian wrote documenting the build-up towards the coming storm (later to be known by the Allies as the third Battle of the Somme), he graphically described his experiences of the combat:

In March 1918, virtually on the eve of the formation of the R.A.F., the situation in Flanders looked grim. The Germans had broken through on the Somme. Big concentrations of troops had been massing for weeks for what was to be Gerry's final attempt to crack the allied line.

The order went out that the German advance must be held at any cost, that the massing support troops must be broken up before the main offensive began. Later it was recorded that on the Western Front the day of the great battle, March 26th marked the birth of Britain's new Air Force, although officially the date of the change-over from R.F.C and R.N.A.S. to the Royal Air Force was to be recognised as 1st April.

However, the British pilots standing by for action didn't much care at that moment what they were to be called. They had just received a momentous order from Major General Salmond, the G.O.C. of the R.F.C. in France, a man not given to over-dramatising a situation: 'Bomb and shoot up everything you can see. Very low flying essential. All risks to be taken. URGENT.'

Every available aircraft took off at dawn. The weather was poor – hail and snow. Flying at 5,000 feet, the planes roared and bounced their way eastwards. Puffs of smoke in the distance told them they were nearing the front line. Occasionally a bigger explosion would send out flames and the debris seemed to settle with impossible slowness. Then the aircraft were passing over the remains of trenches. Dark shapes lay like bubbles in muddy puddles. Some of them moved. Then came the red tracer and the ack-ack, or Archie as the German anti-aircraft guns were called. Still the swarm of bi-planes bore down on the German lines. Small-arms fire reverberated and bombs thudded down into the masses of grey uniforms at point-blank range.

A bugler of the 8th German Grenadier Regiment recorded in his diary: 'Several Tommies flew so low that their wheels touched the ground.'

In fact, Company Commander Nedee had to fling himself out, flat on the ground, and then was struck on the back by the wheels of another machine, thus being run over.

When the first attack ended, the British planes battled against the prevailing winds back to their bases where, refuelled and re-armed, they set off again on their nightmare journey.

'Had the enemy attacks succeeded?' they wondered in the messes that night. Confirmation came very soon, when General Sir Hubert Gough, commander of the British 5th Army, telephoned Field Marshal Sir Douglas Haig and told him the enemy attacks were breaking up.

King George V sent to Haig the following message:

I want to express to General Salmond and to all the ranks of the air services of the British Empire in France today my gratification at their splendid achievements in this great battle. I am proud to be their Colonel-in-Chief.

This operation had indeed been a success for the Allies in holding back the might of the German attacks, for now at least, but at what cost? Tens of thousands of casualties and fatalities were suffered by both sides. The average life expectancy of British airmen was now down to just eleven days ... and there were more phases of this battle to come.

However, now that this battle was over, the fierce pride and individuality of RFC pilots came to the fore regarding the changeover to RAF. Many were reluctant to dispense with their familiar regimental uniforms. Lieutenant William Atkin from Lanark was just one of many protestors when he refused at first to wear his pilot's wings with anything less than his RFC Highlander's tartan trews, or to fly anywhere without his Glengarry, which he kept in a corner of his cockpit.

Chapter Five

1918

Cock Squadron

I n April 1918, just a few days after the battle, Brian, aged 22, was called back to Britain to take command of a Bristol Fighter squadron at Biggin Hill, in defence of London. At this time, 141 Squadron covered the North Kent sector from Biggin Hill down the Thames, guarding the southern approaches to the city.

This airfield had only been inaugurated in 1917, a few months before Brian arrived. Its first commander, Major Babington, had been badly injured in a crash-landing one night and was not able to return. It seems that his replacement was 'hand-picked' for this important task, hence the appointment of Major Brian Edmund Baker, DSO, MC. Graham Wallace, in his book *R.A.F. Biggin Hill*, describes Brian as 'one of the avowed opponents of the Richthofen Circus, Baker was a great leader and an immensely popular commander. Under him, 141 Squadron matured into a highly efficient and very happy unit.'

Not only was Biggin Hill founded as an aerodrome, it was also an RAF wireless research establishment. Their buildings were separate, but they worked together, helping each other.

At the beginning of the First World War, once a pilot was in the air he had absolutely no way of contacting anyone on the ground, including HQ, or any of his comrades in the air. This was a problem that somehow had to be addressed, so the boffins put their heads together and came up with a plan. In their minds, the most important messages would be those from their HQ directing the pilot where to go or what to avoid. Secondly, they wanted to know where every aircraft was so that they could plot their whereabouts at any given time. Of course, they focused on what would be most useful to them – information from the pilots. Orders went out for every plane's observer (if it had one), to plot and mark on a map their position, plus any written reports they might have. These were put into a weighted bag and dropped by the pilot, swooping as low and as close as possible to HQ. Whilst this was useful in some respects, there was no way to relay it to other pilots or ground crews. It was a good try for its time; perhaps better than nothing. However, the pilots were understandably nervous about revealing their positions in case their bag fell into enemy hands, which rather scuppered the whole scheme. They needed something more sophisticated.

Now, at Biggin Hill, that time had come. With all Brian's flying experience, he was keen to work with the scientists and engineers by explaining what was required and by suggesting any design detail changes. Brian enjoyed being their guinea pig, undertaking experiments for them, such as flying different distances, heights and areas, to help them find ways to reduce interference. Most importantly of all, Brian and his men trialled prototypes in their planes on real sorties and reported their problems and ideas for improvements. He wrote of this endeavour:

The squadron was just being equipped with a new advancement called wireless telephony in an attempt to help out our pilots find any raiding German Gotha bombers after dark. These twin-engined aircraft had superseded the Zeppelins, which had proved too vulnerable.

The Gothas came over on day and night raids. Then the daylight raids began proving too costly for them (in both aircraft and lives), so they resorted almost exclusively to night flying. This was the menace the wireless telephone was designed to combat. If the searchlights were unable to illuminate any of the bombers, an intercepting fighter pilot's only hope was to be directed by wireless telephony from the ground to the area in which the raiders were operating.

Brian was a man who enjoyed a good lunch and he soon found just the place for a break between flights. In his own words:

> One of our favourite landing grounds was down the railway line, near a village called Staplehurst. We first found it because the field was designated as an 'emergency landing' field. The main reason for its popularity was that the landlord at the Railway Hotel laid on a very good lunch for us. We got away with it for quite a long time, until one day the Commander of the Wing had some engine trouble during a flight and happened to come down for an emergency landing on our unofficial airfield.
>
> First thing he saw was a whole lot of Brisfits [Bristol Fighters] sitting unattended on the ground.
>
> Not unnaturally, he asked an N.C.O. where they all came from. The sergeant told him we popped in regularly, usually in larger numbers.
>
> After that episode, lunch at the Railway Hotel was definitely off the menu. We had to dine at the Mess at Biggin Hill after that. Rissoles and 'Zaps on a cloud' would have to suffice.

In the days between raids, Brian wanted to keep everyone fit and keen, so he encouraged the men not to just sit around, but rather to take part in things. So he put on a variety of physical activities, starting with daily rugger matches, cross-country runs and paper chases. Brian's favourite sport was the one he excelled in most, namely cricket, so he asked around but didn't manage to locate a full set of cricket equipment. Not a man to be stumped, he got in his Brisfit and flew to Hertford.

His old school magazine *The Haileyburian* takes up the story: 'Captain B.E. Baker DSO, MC, the other day sent a request from the air for some cricket requisites for his squadron. The note fell from a clear blue sky, just as the Quad was full after 3rd lesson.'

Brian knew the school's routines, which hadn't changed, so he timed his arrival just right. He hovered to drop his note, which fell amidst a group of boys. With whoops and cheers all around, one boy picked up and read the note, which asked to borrow some cricket equipment for a few hours. The boy and a couple of his classmates ran into the school, waving the note from the sky.

Brian circled over the Quad and shortly the boys came back out with a master, their arms full of cricket equipment. Landing on the school field, Brian just had time to thank them all before the bell called the boys back in for their next lesson. He loaded the kits into his plane and off he flew back to Biggin Hill for an afternoon of cricket with his lads.

Being based in England didn't stop Brian from writing home, and also allowed for the occasional diversion to fly over his home, sometimes even spotting and waving to his sister or mother in their garden. On one such occasion, a fine day when he had planned to land on the golf course near their house, he had to abort the flight, as he explained afterwards in a letter to his sister:

> Sure enough, I was on the Bristol, but my engine nearly petered out after that dive, so I thought I'd better get as far away as possible before it snapped altogether. I didn't see Lawson's machine [aircraft] on the golf links, but I did see the most awful crush on the lawn. I thought it must be the Vicar's tennis party.

On 4 May 1918, Brian suggested bringing a fleet of Bristol Fighters to fly in formation overhead, 'as soon as we have a fairly clear day'. Sadly, that didn't happen as the night raids accelerated and 141 Squadron had to sleep most of their days away.

Two days later, on a very clear night, the enemy came over in large numbers: 'It looks like business tonight. We were all turned out last night at midnight. We did curse the wily Bosche. But about 20 minutes later is was all clear.'

Another night:

> The radar gave a warning on the old siren, so up we went. Then we had the 'all clear' about an hour later. We heard it was an old sea-plane of ours, wandering about the coast for a forced landing. Wasn't that fine?

Brian's next letter to his sister was fatter than usual because it enclosed a piece of fabric:

It's from Richthofen's machine. It was given to me by old Gould, our armaments officer in 48 [Squadron], who came down to see me the other day. A piece about one foot square was refused at the Stock Exchange for £200, so hang on to it.

Flying continued with day and night raids to repel the Huns and their bombs, but on 19 May 1918 came the last big night raid on London. About forty Gothas flew up the Thames and across North Kent, but the majority of enemy aircraft were repelled in the defence of London. Brian described one such victory:

We went up against them and one of our aircraft, piloted by Lieutenant Turner, shot down a Gotha, for which he and his observer got the DFC. The Squadron took as souvenirs one of the aircraft's guns, a propeller and a black Maltese cross, cut from the fuselage. This victory earned the squadron a congratulatory telegram from the Mayor of London.

After that hectic night, we continued with our wireless control practices, using an old German bomber which had been captured.

It carried six and we'd fly it round the searchlights and the Royal Engineers' listening posts to give them practice with twin-engined aircraft. I was flying it one day with some of the lads in the back, ostensibly showing them the ropes. It must have been dreadfully boring for them and, unknown to me, they had started a rummy school to pass the time.

It came to an abrupt end when one of the propellers flew off. There was a tremendous flap as I wrestled to control the brute, which had gone into a spin. I did so long enough to land in a field in Essex, without anybody being hurt.

In a letter to Win dated 27 August 1918, Brian wrote an initially light-hearted letter, which changed tone at the end: 'I can't think of anything more to say at present. A squadron was completely written off with bombs the other night and had 13 officers knocked out. Wasn't that jolly of the Bosche?' One can only imagine the impact that must have had on Brian and all the pilots in Home Defence.

When the weather conditions were advantageous to the Huns, all the men of 141 Squadron had to be on standby, ready to do battle and head off their raids. However, on murky evenings, with not much else to do, the officers' mess inevitably became a focus for drinking, leading to rowdy sessions for some of the lads almost every night. One of the initiatives Brian thought up was to provide a book and ask all the officers to write in it any suggestions of ways to improve the comfort and amenities of the mess. A few days later, he opened the book after dinner and started to read them out … 'Cannot those everlasting rissoles for breakfast be varied? Signed R. Suppards.'

After moment's pause, Brian looked up and around the room, almost certainly noticing the anticipation on their faces.

'Who is R. Suppards?' Brian asked, which was the cue for gales of laughter across the room, followed by a jolly recitation of the 'Epitaph on the vault of Richard Suppards':

A pious mortal and a wise,
Who never cheated or told lies,
Beneath this marble tablet lies,
R. Suppards

Beloved by all the friends he knew,
He kept an inn and wealthy grew,
His sign, an angel, grandly flew,
R. Suppards

Great store of friends he had but that
He loved the most his tabby cat,

He kissed and fondled as she sat,
R. Suppards

But sadly Richard's life was ended,
By falling from a beam he mended,
Upon a fence he hung suspended,
R. Suppards

They took him down, in white arrayed him,
They shaved his head and tidy made him,
And sadly in the grass they laid him,
R. Suppards

And though his form is laid in night,
And though his bones are out of sight,
His virtue still will come to light,
R. Suppards.

The generally good-natured rowdiness in the officers' mess continued but it wasn't long before a few of the drunken lads broke some windows. This came to the attention of Wing HQ, who imposed fines of one shilling and sixpence for every broken pane of glass. That enraged a young Canadian lieutenant so much that the following evening he went round chanting his displeasure with the fines as he kicked out every remaining window pane. Meanwhile, some of the others who had incurred fines opened the lid of the hired grand piano and started plonking its keys so loudly that some of the strings became discordant. Not liking the sound, they carried the piano outside and disembowelled it.

The following morning, Brian phoned his 'spy' at HQ and discovered that somehow word of the smashed windows and the wilful damage to the piano had reached the ears of the general and he was going to pay a surprise visit to Biggin Hill that afternoon. Brian immediately assembled all the now sober officers together and announced the impending visitation. They had just a couple of hours to remove all the window panes from the cooking and service areas and affix them in all the gaps around the mess.

An inspection of the piano made it obvious that its innards were beyond repair, so, with a bit of quick thinking, Brian deputed the remaining officers to bring the piano back into the mess and polish it till it shone. Then anyone with framed photos in their quarters brought them through to arrange tastefully on top of the piano, with a large photo of the King in pride of place. Finally, Brian stood back and is reported to have said: 'That might prevent the great man from looking inside. Let's pray he doesn't touch the keys.'

Within the hour, the general arrived and Brian took him on a tour of the airfield's facilities and aircraft, then inside and through to the officers' mess. Here the general made a proverbial beeline for the piano. Brian stepped forward to distract him by pointing out the military photos and of course the ornately framed photo of the King. The general's hand reached out as if to open the keyboard, with no doubt several officers' collective intake of breath … then seemed to change his mind, as if perhaps thinking he must have been ill-informed. Congratulating Brian and his senior officers on how well everything looked, he took his leave, to the great relief of all.

With the comparative lull in enemy raids, Brian's brigade commander decided he needed to think up an idea to occupy and challenge the pilots and aircrews to raise their spirits. So he organised a 'Squadron at Arms' competition. Points would be awarded for formation flying, wireless telephony, gunnery, aerobatics, and the general appearance and upkeep of airfields and station buildings. The winning squadron would receive a cup and the title of 'Cock Squadron of the Home Defence'. In addition, they would be allowed to display a fighting cock emblem on the fuselages of all their aircraft.

This competition enthused Brian and he immediately planned what 141 needed to do to win the cup. He knew his men would do very well on most of the elements, especially wireless telephony and the flying challenges, but as he walked around the airfield he realised how bare and unkempt it looked. The buildings appeared rather boring too. He gathered all the lads together and told them about the challenge, hoping they wouldn't feel it was too big a task. But he needn't have worried. As he later wrote:

> The squadron had a large percentage of Canadians and as soon as they heard about it, they were determined to win the title. Believe me, when Canadians get going, they take a lot of stopping. To make sure the squadron buildings and surrounding areas were up to scratch, they set about doing a lot of very unorthodox gardening. This consisted of importing (i.e. borrowing) and planting young trees and bushes with no roots and plants in pots.
>
> By the time they'd finished their instant horticulture, the whole place looked like Kew Gardens in miniature. The inspecting officer was amazed to see such a magnificent display.
>
> When the wireless telephony part of the show came along, we were practically 100 per cent. Thanks to our training, the other squadrons reckoned this was a swindle. Our opponents were convinced we knew in advance what was coming ... and there may have been some justification for this. Knowing the boys under me, I wouldn't have put anything past them. Suffice to say, I knew nothing about it ... officially.
>
> Anyway, we won. The silver cup and a live game-cock were presented to me by the then Air Minister, Lord Weir.

Brian's 141 Cock Squadron had already gained notoriety for its exploits in the air, but now its reputation and its proximity to London made Biggin Hill the place to take VIPs. So it was that in the autumn of 1918, Buckingham Palace arranged a visit to Biggin Hill by a diminutive Japanese royal – Prince Hirohito of Higashi Fushimi. The weather that day was atrocious, so most of the planned events had to be cancelled. Brian greeted the prince and escorted him round the station. He showed a particular interest in the Handley Page O/400 when Brian explained that this aircraft had made the first wireless flight, to Paris and back.

Finally, despite the downpour, the prince turned up his coat collar as Brian took him outside to watch the squadron's flypast, with their red-and-gold cock crests gleaming through the rain.

As Brian saw off the prince's cavalcade, he would have had no idea that two decades later, this seemingly harmless little man would be his enemy.

Chapter Six

1918–1920

Peace at Last

In the early morning of 11 November 1918, a drowsy operator at Aperfield Court, the RAF's wireless testing station at Biggin Hill, who had nothing better to do tuned into the Eiffel Tower station and picked up a message, which shook him wide awake:

Marshal Foch to Commanders-in-Chief
Hostilities will cease on the whole Front as from November 11th at 11 O'Clock (French time).

The excited operator went straight to tell Colonel Blandy, who in turn passed the wonderful news to Major Brian Baker as well as all the local padres and vicars. Within a few minutes, the bells of Cudham and Westerham pealed merrily for all to hear. These churches thus gained the distinction of being the first in Britain to proclaim the longed-for peace. John Westacott heard the bells from Cudham Lodge, mounted his horse and galloped across to the north camp of Biggin Hill, sounding his view-halloo horn all the way. By now, all the local residents were out on the streets, cheering and dancing. The pilots brought out their planes and gave joy rides to all the young girls.

Sometime afterwards, Brian was asked to recount his memories of Armistice Day for a BBC radio broadcast:

On armistice day in 1918 I was a member of 141 Squadron at Biggin Hill, a celebrated Bristol Fighter squadron which had just won the Cock Squadron Trophy, from among the squadrons engaged in the defence of London.

We had a number of Canadian Pilots and Air Gunners who were first-class fighters and also enjoyed a bit of fun.

Early in the morning of 11th November the alarm sounded, which roused us from our beds and we ran down to the hangars where the airmen were rushing out the duty flight of Bristol Fighters. On arrival on the tarmac we were informed by the Commanding Officer that a signal had just been received that at 11 O'clock that morning an armistice would be signed and the war would be over.

There was a great cheer and we all returned to the Mess for a Champagne breakfast.

While we were celebrating, Mr Westacott, the farmer from down the lane, rode his horse into the Officers' Mess then rode round and round the billiards table, sounding his 'View Halloo' horn.

After breakfast, most of the squadron disappeared to London. On arrival in the city, there were vast cheering crowds everywhere and I remember seeing a Crossley Tender coming down the Strand, firing Very lights out of the back. During the morning, I saw a heavily laden taxi, stuck in the revolving door of the Criterion Restaurant. I found out afterwards that it had caught fire.

Later we went to the Savoy Hotel, where the chaos in the dining room had to be seen to be believed. Fortunately for the owners of the Savoy, there was a wealthy man sitting on the balcony, enjoying the good-natured confusion below and he asked the manager not to interfere, as he wold gladly pay for all the damage that was being done ... and I believe he kept his word.

Later in the evening, the final episode took place when a fire was started at the foot of Nelson's column in Trafalgar Square. Large heaps of wooden blocks were scattered all around for road repair purposes, which were good fuel. These were all piled up in a heap at the base of the plinth facing Cockspur Street. The blaze was considerably helped on by a German gun from the Mall. The Fire Brigade arrived and their jack-knives soon produced the finest fountain display that Trafalgar Square had ever seen. Any doubt about the authenticity of this fire can be dispelled this day by looking where there is a burnt crack in the stone in Nelson's column, which is still plainly visible.

We returned by train to Bromley, where reports of our efforts had preceded us and we were escorted from the train to the Police Station. After an exchange of head-gear, they kindly gave us coffee and rang up Biggin Hill Transport to come and collect us.

In fact, it is perhaps not surprising that Brian didn't recall or wish to relate every detail of that day, and of course there were some differing accounts from disgruntled residents, especially those of the Savoy, who were recorded by Graham Wallace as being 'scandalised when a party from Biggin Hill burst in, whooping like foreigners and demanding Champagne at ten in the morning'.

There were other accounts too of the incident with the taxi and also their exuberant hauling of the Hyde Park guns up the Mall and parking them under Admiralty Arch. However, it is to their credit that their high spirits did no harm to anybody, although they themselves might have felt the worse for wear by the time they set out for home.

Two or three days later, Brian received a disgruntled letter of complaint in a flimsy purple-lined envelope:

The Royal Bell Hotel,
Bromley, Kent

Dear Sir,

It is with a certain amount of regret that I feel constrained to write to you to ask you if it is possible to assist. You will see by the above address that I am at present staying at Bromley, and at the above-mentioned Hotel there are dances practically every night, patronised by Officers of H.M. forces and their Lady Friends, usually until 03.00.

With the dances I have no fault to find, believing as I do that everyone is at liberty to find their own outlet of enjoyment, but

I do most strongly object to these gentlemen and their friends coming into the private parts of the Hotel and sitting about in the corridors and landings, making an infernal row, which disturbs my night's sleep.

The last two nights have been absolutely awful. On Monday night, or rather Tuesday morning, I felt compelled to go out and asked them to be quiet, pointing out that I (and others) had to be up next morning and ready for business. I was met by the following 'GoTo ****'

Last night, I fancy, was a farewell dance of the R.A.F. 14 1 Wing, and at about 1.00, about a dozen or so Officers were marching up and down the corridor, singing, or perhaps I had rather say shouting 'R.A.D. 141 RAF. I41'. I think Sir you will agree I have some ground for complaint.

My second complaint and in some respects perhaps the most glaring is the number of cars, of government cars that are used at these dances does appear to me to be a frightful waste. These cars last night were kept running from about 2 to 3 O'clock (just to keep them warm I suppose). The noise and fumes in my bedroom were simply awful. So, what with the merry dancers in the corridor and the back-firing in the street, just under my window, I feel compelled to write this letter.

I do sincerely hope that you have a spare A.P.M. who will undertake to make the necessary enquiries, but of course it will be useless calling during the day.

Last night I counted fourteen service cars and tenders.

Yours Faithfully,

A.W. Matone

Once all the celebrations were over, Brian and his contemporaries realised that the powers that be were now beginning to dispense with both officers and all ranks in each of the forces. Brian could be no less sure than anyone else of maintaining his position in the RAF. These were worrying times, when all he had ever wanted to do was to fly.

Writing a piece for *The Haileyburian* after the war had drawn to a close, Brian explained:

> At the end of the war I was a Squadron Commander. I realised there was no hope of retaining that rank – neither did I mind very much. The grand thing was to be able to stay in the R.A.F. with a permanent commission.

On New Year's Day 1919, Brian was surprised and pleased to find that he had been awarded the AFC (Air Force Cross).

On 11 March 1919, 141 Squadron moved en masse to Tallaght Aerodrome, just outside Dublin, when Southern Ireland was still part of Great Britain ... but not for long, if the Irish Republicans had their way. In fact, following a landslide victory for the Sinn Fein party in the Irish elections, they had formed a breakaway government and declared Irish independence on 1 January that year. This was a centre of unrest, to say the least, and to make things worse, the British were now busy building two new aerodromes on the outskirts of Dublin: Tallaght and Baldonnell.

Brian's 141 Squadron moved into their new quarters, sussing out the airfield and checking that their aircraft were safely ensconced in their new hangars. It must have been a busy time for them, as Brian didn't write to his sister until three weeks later.

> 141 Squadron – arrived at new aerodrome in a terrible mess.
> Tallaght near Dublin
> 22nd March 1919
>
> My dear Win,
> This place is a most awful mess and will take weeks to get straightened out. Mercifully there is another squadron coming and the C.O. of that is senior to me, so he will be station commander and will have to take over all the buildings etc. so I shan't be

responsible for them. Just my own squadron and nothing very much is very nice.

He'll cuss a bit when he sees what he is in for, I know.

I am just off to have a look around Dublin. I haven't been outside the camp since I arrived. It is seven miles from anywhere.

Seven machines have not arrived, but the weather has been too awful for words and finished up today by snowing hard. It is beautifully fine snow. I do hope it's going to clear.

There was a social on one of the other aerodromes a few nights ago and they had all their machines tied up and their rifles and ammunition pinched from the guard room. All ours is now surrounded by barbed wire and guarded by 30 men.

So it ought to be quite safe.

No more news at present.

Love to all.

Affec.

Brian

His next letter, in August, was full of all the sports fixtures he had had and was about to have, including cricket, of course, plus sailing, an aerial Derby Day, Tug-of-War, Commonwealth sports and Saturday sports, where his team won the general's Challenge Cup. 'It was all a jolly good show.'

In November 1919, Brian and his squadron were moved to Baldonnell Aerodrome, not far from Tallaght and still close to Dublin, where the troubles had escalated. The day-to-day routines, social and sporting activities continued relatively peacefully at and around Baldonnell Aerodrome for several months, alongside their regular flying patrols, which were their main job, watching out for disturbances.

On 25 November 1920, Brian's letter to Win described how explosively things in Ireland could change:

We've had the most terrific week of it. Fenians [Irish Republicans] laid out 14 officers on Sunday morning in Dublin. So everyone was confined to camp. As a matter of fact, we still are. I went up [in a plane] in the afternoon to have a look round and saw a crowd of people rushing off to Croke Park. A huge beat-up ensued and several people got shot. I was right over the show but never knew it was going on. Rumours went round that we fired on the crowd and a question was asked in the House to that effect We hadn't got even a pea-shooter on board, far less a gun.

These excitements carried on all the next day and every minute we expected to hear that the Black and Tans [British recruits] were wrecking Dublin, but nothing has happened so far.

Curfew is now at 10 o'clock and a strong guard put on, not so much to keep us in as to keep the Black and Tans out, as far as I can see.

The General was here today and I led a formation over the show for three and a half hours. We were absolutely finished when that was over.

Black and Tans followed the procession and all the hats and caps that were not taken off were thrown in the Liffey.

There were several fires all around tonight, so perhaps they are getting busy ...

I was in Dublin this evening with Turner. All seems quiet now. Two more people were shot this morning, but I don't know what side they were on yet. Some of our officers were in one of the hotels that were raided on Sunday and heard the Fenians asking if there were more officers there. On learning there were only R.A.F. officers staying there, the Fenians said they didn't want R.A.F., only infantry, which was rather a comfort, especially for the fellows in the hotel.

I suspect there will be further developments in the next few days.

I hope the vicar's corns will soon be better.

In 1920, amid the violence and chaos that was Ireland, British control collapsed and No. 141 Squadron was disbanded.

Sad as it was, after working together so well and for so long with his squadron family, Brian took this as a positive opportunity to do something he had been considering for a while.

He applied and was accepted to train as a flying instructor at the RAF's Central Flying Training School (aka RAF Aircraft Engineering School). Brian enjoyed his year there and in January 1921, he took on his new role of instructor at the RAF's School of Technical Training based at RAF Halton. No doubt, his experience with wireless telephony on the field and in the air during combat must have been invaluable.

In December 1921, Brian was assigned as an instructor to No. 4 Flying Training School at Abu Sueir in Egypt.

Chapter Seven

1922–1925
King Tut's Chariot

On his arrival in Egypt, Brian couldn't wait to write down his first impressions in a letter to his sister:

A.F.S., R.A.F, Anu Sueir,
Now at Shepheard's Hotel, Cairo
15th December 1921

Dear Win,

We have arrived at Abu Sueir, which is rather a desolate place, but have come down here for a few days. This place is wonderful – The best hotel I've ever seen in the world. It costs 25 shillings a day, which includes everything. Wonderful rooms, food and dancing.

The country coming down from the Canal was very interesting. All desert for about a third of the way, with an old camel or two lumbering slowly across it. Then we came to the concentrated part of it and crowds of people with camels and donkeys working in the fields.

Prize dates and tangerines seemed to be the most plentiful things growing. Wonderful-looking birds, including sparrows of course, white herons, ravens, huge white hawks, a very huge, speckled kind of kingfisher and I also saw an ordinary kingfisher.

We are going to get the kitchens developed this afternoon and are going to the Pyramids tomorrow.

The new course does not start till February. So we shall probably get a fortnight's leave on base. That being the case we are going up to Luxor and the Line.

The Christmas mail leaves today, so I must rush with this off to the post.

Merry Christmas to you all,

Love to Mother,

Yours Affec.

Brian

Part of the RAF's expansion programme, the Flying Training School at Abu Sueir was a brand-new establishment, opened in 1921, so work was ongoing to expand its facilities for four instructors and the first full cohort of RAF students. Brian was one of its four inaugural instructors. Once the campus was ready, in a few weeks, he would be training pilots for squadrons across the Middle East.

Initially equipped with Avro 504s, the school's main purpose was to provide training in all aspects of flight, including co-operation with the army. The long-term goal was to increase first-line strength across the region from fifty-two to seventy-two squadrons.

As the campus wasn't ready for use yet, Brian and his colleagues were temporarily living at the hotel and making good use of their spare time, as shown in extracts from his next letter home:

Shepheard's Hotel

Cairo

22nd January 1922

My dear Win,

There was a splendid dance here last night, not quite as crowded as usual. There were quite a few people from our boat still going strong, so we had a very cheery time. ...

It is getting much hotter now in the middle of the day but the wind has kept it down a bit.

We are going round to the museum tomorrow to have a look at Rameses I. He's quite OK, so I'm told, despite his 5,000-odd years.

Yesterday we went to the Sporting Club. Quite good looking – any number of tennis courts, a golf course, cricket, polo, etc. Also a wonderful view of the Mohammed mosque with the setting sun shining on it – a wonderful sight.

I am going to try to get down to the Mosque tomorrow and the native bazaar. You can pick up wonderful bargains there. …

I met one of my old flight here last night named Binnie. He was with me in 1917 and is now doing aerial mail, from Cairo to Baghdad.

It's a most extraordinary place for meeting people and seems to be a rendez-vous for the World.

In his next letter, Brian described playing in a hockey match, in which, 'I got a crack on the wrist, which put it out of working order for a bit, but I got a couple of goals which won the match, so it was some consolation.'

At one of the many dances, Brian was in good company, with Field Marshal Allenby and Lady Allenby, as well as 'a crowd of others' he knew – 'about 340 people, all in mess-kit and it all looked rather nice'.

The following day he did some shopping at the mosque, which impressed him greatly, as he described to his sister:

I've never seen anything quite like it The street scene is like 'The Garden of Allah', but much narrower.

The Mosque is an extraordinary place. I've never seen anything like it. Men of all nations sit in little alcoves with their goods spread out in front of them.

We first went into a scent shop and were invited to partake of coffee, which we did. I think it was the finest coffee that had ever been made. Then we smoked his cigarettes.

The flower scents were marvellous. I paid him to send you two bottles.

Brian carried on round, buying furniture and rugs and a copper tray for his hut. 'They had an awful sand-storm here while we were away.

It must have been awfully strong because it removed one of the walls of the new hangars which were in progress of being built.'

On the following day at Abu Sueir, Sunday, the sandstorms had worsened, as he said in his letter of 31 January 1922:

I don't know whether you got my last letter because the mail seems to have gone to pot the last ten days. We haven't had any letters for ever so long. It's probably been very rough, and we have had two days of terrific wind and yesterday a sand-storm from 9.30 in the morning to about 4.00 in the afternoon. Sat around all day and slept all night. ... Tomorrow I am going down to Ismailia to play tennis with Squadron Leader Fiennes.

In between all of Brian's social, sporting and sightseeing activities, he did occasionally mention the work he was in Abu Sueir to do – mainly training RAF recruits and extending his knowledge of technical developments in flight and related disciplines. In mid-February, Brian explained in his letter to sister Win about his growing fascination in one particular aspect:

On Saturday I took over the Meteorological department of the station. I've been down there all morning with the Sergeant Major in charge, trying to learn something about it. I discovered a large telescope which pleased me very much, as I can now look around the stars, which of course are simply magnificent out there, but to me seem to be all in the wrong places. The Pole star is quite low down and Sirius is nearly over our heads.

We had the devil of a Kansene [sandstorm] yesterday morning. The wind was 40mph (one of my instruments recorded that, so it must be true). The sand was so thick that we could but see 20 yards. We had a machine out in the dust, attempting a forced landing and another beside it, trying to get him off again. The wind just picked them both up, twiddled them round and smashed them into each other.

The weather improved, so Brian and a mixed group of friends went sailing, swimming, fishing and camping a few times at glorious Lake Timsah and Brian was deputed to do the cooking. On one such outing, Abu Sueir's meteorological sergeant major sent a wireless message through to Brian that there was a 55mph wind at 100 feet over the aerodrome. The warning allowed Brian and his friends to move to a safer place.

Over the very hot months, Brian wrote home that he felt like he was in a furnace, but despite that, he managed to play and often win cricket, hockey and tennis matches. When the extreme heat abated, in June 1922, it coincided with Brian's break from work for Whitsun and the King's birthday. Brian's rising interest in all the reports coming out of the Valley of the Kings and elsewhere persuaded him to go to see for himself what was going on. What new treasures might they find? He wrote all about it to his sister:

26th June 1922

My dear Win,
I arrived at Heliopolis, then went to Sacara [Saqqara] and Memphis by car with a party of five. I saw the tomb of the sacred bulls, the Colossus of Rameses II at Memphis and the famous step pyramid, which I thought very dud. We trouped around on camels and numerous photographs were taken, which I will send in due course.

The drive down along the Nile was simply grand and, as there were no tourists, we had the whole place to ourselves. I was quite expert on a camel. It's a most comfortable mode of conveyance when you get used to it.

I have joined the sporting club and we are trooping off there on Saturday to see the Prince of Wales play polo, and then on to Kasr-El-Nil Barracks in the evening, to a boxing show that he is attending.

The course is an excellent show. I arrive at 9.00 a.m. by train from Heliopolis and depart at 2.45, with a half-hour break in the middle for a drink. All the afternoon is either cricket or tennis – simply wonderful.

I am in a tournament on 17th at Heliopolis and have a good partner so we have hopes of doing something.

Last night we went to the Pyramids with a party of 30 to see them by moonlight and on Tuesday, being a full moon, I am going with another crowd. They look much better by night than by day and ten times as large.

We had a cricket match against 9th Lancers last week. I knew most of them out in Ireland and their captain of the Rookery family. I also met a girl at the dance last Friday who knew nearly everybody I know in Bromley.

Seven months later, in January 1923, Brian had a fortnight's leave – long enough to embark with a friend on the trip he had looked forward to since he came to Egypt. In his own words:

We left Cairo at 3.00 in the morning. We had sleepers to Luxor and arrived here at 9.30. We were very stiff, so we decided to have a game of tennis to work it off. This turned out to be a very lucky move as we fell in with an old fellow who had his own river house-boat and two lovely daughters. We also met the *Morning Post* correspondent who knew all about what was coming out of the new tomb the next day. They both played tennis with us all afternoon. Then we were well away for the dance in the evening as the daughters knew all the ladies at the hotel.

At the dance I ran into the Chief of Antiquities in Upper Egypt in the bar. He gave us all passes to visit any tomb, temple or statue in the whole of Upper Egypt for ten days

Next morning we trooped off in a body to the tomb of Tutankhamun. Five on donkeys started off and were afterwards joined by the *Morning Post* man and old man Lewes on donkeys, then the daughters in a kind of car – a most delightful sight.

There were only about a dozen people at the tomb [and] as there wasn't much sign of anything coming out, we went down the tombs of Rameses VI and II. The paintings on the walls are just as good as I'd seen it in pictures and some of the figures were not that difficult to read when you get down to it.

Next, we were just in time to see the chariot wheels being removed from the new tomb [of Tutankhamun]. Everything was covered with gold, on the knobs and the wheels which are studded with gems. Chairs, thrones and some white hair in a box were the next to come out of the tomb, then they packed up for the day.

We all went off to 'Looks Rest House' for lunch.

In the following days, Brian and his party explored almost every other archaeological site and beauty spot in Egypt, punctuating their grand tour with donkey and camel gymkhanas, visits to engineering marvels, ancient mosques and temples, as well as taking part in various sports matches and, of course, the inevitable parties and dances. One night towards the end of his trip, Brian and a few friends went up the Nile in the Lewes's dahabiya (boat lit up with fairy lights), through the orange groves, picking and eating the sweet, succulent fruits.

The sun was setting when we came downstream. It was a glorious sight. The whole show reflected in the water, a deep purple, while the colours in the sky changed every few minutes, reflecting on the slow ripples of the Nile.

Chapter Eight

1924–1932

Experimental

O n 24 February 1924, while still serving on the Directorate of Training, Brian was posted to the Middle East to take command of the Aden Flight. His new base was RAF Khormaksar in Yemen – quite a contrast from Cairo.

Brian flew himself there and his first sight of his new command was a barren, concrete airstrip, with the sea to one side and tall cliffs of jagged rock to the other. It was only when he was on the ground that he noticed the familiar, friendly hangars, and what looked like a village or small town at the foot of the cliffs. He was glad to see nearby not only the hangars but also a number of RAF trucks and a surprising variety of aircraft, gleaming in the relentless sunshine.

Having heard that this airfield had the motto 'Into remote places', he was not surprised by his new surroundings. Nevertheless, as always, he was keen to see it all for himself in the next few days. First, he wanted to meet everyone on the base and then settle into his quarters.

RAF Khormaksar was established in Aden in 1917 and had gradually expanded during the intervening years, as Britain extended its influence across the Arabian Peninsula. Aden was an important British staging post to both Singapore and East Africa. On learning that reinforcements were scheduled to join the Aden Flight before long, Brian started to plan for the necessary expansion of the base to extend mess and living quarters for all the additional pilots and ground crews, not to mention the extra storage for aircraft, fuel and other provisions. Brian planned it all out and the work began, but in April 1925, just a year and two months after he had arrived in Aden, he had to leave it all behind as he was called back to No. 4 Flight Training School in Egypt, this time as both instructor and commander. He was also asked to serve again on the staff of the Directorate. While there, on 1 July 1925, he was promoted to squadron leader. By now, the Flying School had acquired some Bristol Fighters, the latest version of his favourite aircraft.

A few months after his return to Abu Sueir, Brian was on the move again. It was spring 1926 when he left the vibrant colours, sounds and scents of Egypt and travelled northwards, back to 'Blighty', in the aftermath of the General Strike. He enjoyed a brief but very welcome leave with his sister and mother in Hertford, where he met up with old friends, visited his old school, Haileybury, and organised or joined some tennis and cricket parties, plus a game or two of golf.

In late March 1926, Brian took up his next role as commander of the Experimental Section of the Royal Aircraft Establishment at Farnborough Airfield. Of course, this appealed very much to Brian, after all the wireless telephony experiments he had helped with at Biggin Hill. Those days seemed a long way back in his past, when the Great War had been raging and the peace came to save them all.

The Farnborough site had begun in about 1905 as 'The Army Balloon Factory', which was part of the Army School of Ballooning. These were the 'dirigible balloons' or airships. It had also experimented with Cody War Kites.

Around the time of Blériot's first flight across the Channel, when Brian was a young schoolboy, the Royal Aircraft Establishment, with its new designer Geoffrey de Havilland, was experimenting with Britain's early aircraft designs. Now, all these years later, under Brian's command, the engineers were trialling various approaches to manufacturing new types of aircraft. They modelled various shapes for future aeroplanes and tested them. They also began to experiment with newly developed materials.

At this point, it was decided to rename the site the Royal Aircraft Factory, and later, the Royal Aircraft Establishment or, as it became known, the RAE. The factory had always been at the heart of this base, designing and building fighter planes, some of which Brian had flown in both combat and reconnaissance during the First World War. Under Brian's command, there were test pilots who tried out the new prototypes and it seems likely that Brian himself joined them to help troubleshoot and suggest ways to solve problems. He always enjoyed doing that. He also put forward his own ideas for improvements, based on his extensive wartime experiences. Wireless communication had also developed new features, in which Brian took a very keen interest.

For once, perhaps because of his expertise in these aspects, the RAF hierarchy enabled him to continue at the RAE for a considerable length of time. However, inevitably the time came for another move, but his next destination must have brought back many memories. The news came in December 1929 that Brian had been appointed commanding officer of No. 32 Squadron at Biggin Hill. It was some time since he had been there, but he had heard a rumour that things had changed and it was all quite a shambles, but maybe that was a while ago. Well, he would soon see for himself.

After Christmas at home, Brian packed up and drove to Biggin Hill. Perhaps the first thing he noticed was that the shabby old wooden huts, hastily put up during the war, were still there, but looking fragile. He recognised the old hangars too, but where was the officers' mess? It was getting dark, so he would have to find an officer to take him to his quarters, and then find somewhere to eat.

That evening he learned the whole story. The good news was that the Air Ministry had decided to retain Biggin Hill in action as a permanent station. The bad news, Bran's dinner companion told him, was that the officers' mess had burnt down, so they were temporarily in an unused workshop building until a new mess could be built. But that was not all …

The following day, Brian met with a man from the Air Ministry for a full debrief. He learned that the airfield was dangerously restricted for modern aircraft and there was a great lack of accommodation and facilities. There was also a pressing need to double the size of the base, but this would mean acquiring land on a large scale from surrounding property owners. A major reconstruction project was needed, for which plans had been drawn up two years before and agreed in principle. However, two major problems had held everything up. First, there was the funding. The Treasury could not be persuaded to provide the huge sums of money needed to meet the projected costs of reconstruction and land purchases. The second problem was persuading (and paying sufficiently well) surrounding landowners to sell their property to make it all possible.

In its early stages, this scheme seemed to be proving to be almost intractable. Brian looked at the plans and talked to a few people, both in the RAF and in the Air Ministry, urging for action so that decisions could be made. Meanwhile, he looked into which landowners had been contacted and what the outcomes were: 29 acres were required, but the geography of the site's surroundings meant that some had to be small plots, whilst a few could be larger. In fact, there was one farm on the map that could provide most of the land the reconstruction would need. It was Cudham Lodge Farm, whose tenant was John Westacott, the jovial fellow who had ridden his horse round and round the billiards table in the old officers' mess on the morning the Armistice was declared. Brian soon learned that he had already vacated his farm and the owner had sold the land as requested.

Whilst all this was necessarily of interest to Brian, in order to provide better for his men, he was of course also flying and testing aircraft himself and fulfilling all his responsibilities. It therefore must have come as a great surprise when he received a message out of the blue that the funding had come through for the full reconstruction to go ahead.

The contractors turned up in the summer, just six months after Brian's arrival, and made a good start, but it would take three years for all the work to be completed. Meanwhile, the only aircraft that could safely operate from Biggin Hill were those of the Night Flying Flight, which were mainly used by the army's anti-aircraft school in the South Camp, and by the long-standing Wireless Experimental Establishment for their communications work. The rest of Biggin Hill's planes would move to Kenley.

As usual, whenever he had some leave, Brian spent most of his time playing cricket or tennis and sometimes hockey or other sports. Indeed, he played in both the cricket and hockey teams for the RAF. Ever since he had scored 139 runs in a cricket match between Haileybury and another public school and soon after, at the age of 18 played for Hertfordshire, he had continued to excel at cricket. So perhaps it was no great surprise that he was selected to captain the RAF cricket team against the navy at Lord's. On that occasion, they drew, but he continued to play competitive cricket at the highest level whenever he could.

On that day at Lord's, the youngest member of Brian's RAF team was a brash young pilot by the name of Douglas Bader. Despite the fourteen-year difference between them, they struck up a friendship and discovered that Bader was in 23 Squadron, which had only recently moved from Biggin Hill along with most of the aircraft. One evening, Brian was invited to a dinner dance at Reading Avro Club with his fiancée, Jaimsie, whom he had met on a romantic cruise on the Nile. They were enjoying drinks on the lawn with others, when they heard the unmistakable sound of a Bulldog Mk. IIA, K-1676 approaching. Within seconds, it flew into view, performing low-flying aerobatics across the adjoining field. The pilot dipped his wing, as if in greeting, but too low – recklessly low – and crashed. Brian immediately ran over to see if he could help.

What he saw when he looked into the cockpit came as a great shock – a very badly injured airman, whom he recognised as his reckless young cricket friend, Douglas Bader. An ambulance was called and Bader survived, but he had to have both legs amputated. It would take a long time, enormous determination and two new legs before he could fly again.

Finally, in the summer of 1932, the reconstruction work at Biggin Hill was completed and everything was ready to open up and be in full use again, with all its new and fully refurbished accommodation and facilities. These included, for the first time at Biggin Hill, some married quarters. There was even a rose terrace outside the smart new officers' mess. Finally, the pilots brought back some of their aircraft from Kenley. These planes were now augmented with some Bulldogs and the new Hawker Demons to try out.

Now there was just one test to pass: The Inspection Visit. A high-ranking officer from Air Defence – an air vice-marshal no less, sent a message to the CO, Wing Commander Dacre, announcing the date of the inspection and a request that all areas be open for the inspector to see. This was a tricky request, but Brian and his fellow officers reckoned they could redirect the man for his own safety.

Inspection day dawned and the air vice-marshal arrived, dead on time. Everywhere had been smartened up with borrowed new bedding and furnishings. The inspector looked through all the open doorways, into cupboards and drawers. He then insisted on going up to the top floor of one of the new blocks … and that's where things could have gone tragically wrong. Brian could not deter him so they reluctantly followed him up the stairs, where he marched towards the first door and turned the shiny brass handle, but it wouldn't budge. The key was sent for and it quickly arrived. The air vice-marshal grabbed the key and turned it in the lock, but the door still would not open. He pushed and pulled it and shook it, to no avail. His previously genial manner turned to anger.

'My orders were perfectly clear,' he growled. 'All areas must be accessible for this inspection.' The colour rose in his face as he shouted, 'Get me the right key.'

At this point, Brian was relieved to see Wing Commander Dacre joining the inspection party and taking over.

'I must apologise wholeheartedly, Sir, that we haven't yet checked the locks.' He turned towards the door in question and gave it a try.

'I do apologise, Sir,' he said. 'I'm afraid this one must be faulty.'

'Hmm,' shrugged the air vice-marshal, as he reluctantly turned to go back down the stairs.

It was a close escape. Had the inspector been able to open the door he could have fallen through the hole to his death. Fortunately, over an excellent lunch in the officers' mess, he calmed down and eventually gave a good report. He never discovered that the money had run out before the completion of that section, and that the absence of the key had saved his life.

Brian and his 32 Squadron must have had a good summer that year, enjoying their new surroundings. However, after three years at Biggin Hill, he received the news that he was on the move again, and this time it was to the other end of the country.

Chapter Nine

1932–1937

Training with Catapults

On 1 January 1932, Brian was promoted to wing commander and took on the role of chief instructor at RAF Leuchars, near St Andrews on the east coast of Scotland. This news was swiftly followed by an embossed invitation from King George V and Queen Mary to Brian and his Scottish wife, Jaimsie, to join them at their forthcoming Royal Garden Party at Holyrood House in Edinburgh. This was Brian's first such occasion.

RAF Leuchars had originated in 1911 as a field with a balloon squadron and a training camp built in the adjoining Tentsmuir Forest by a squad of Royal Engineers. The first aviators who used this field were in flimsy aircraft, 'held together with sealing wax and string', as the engineers described them. One of the main reasons for the early popularity of this airfield was its proximity to the garage at St Andrews that used to sell cans of petrol. Indeed, this was the very same mechanic who had sold a can of fuel to Brian when, in 1915, he performed a practice landing on St Andrews Sands.

Knowing this area so well from those days of his own flight training, Brian must have felt very much at home at Leuchars, giving back to Scotland what it gave him at the almost neighbouring Montrose airfield. Leuchars had always been intended for a training facility and its first title was the 'Temporary Mobilisation Station'. Its intake comprised keen young lads with a yen to sail above the ground. Of course, they soon realised that it wasn't as simple as that.

As chief instructor in 1932, Brian was tasked with two key responsibilities. The first was to train new young pilots in all the skills of flying, along with such additional aspects as navigation, communications and artillery. With the possibility of war on the horizon, Brian's other and more pressing challenge was to train more experienced pilots how to fly their planes on and off aircraft carriers, thereby converting Phantom pilots to carrier operations, in high winds and seas. First he would have to learn for himself how that could be done.

Brian wrote a newspaper article about his experiences in educating himself – and collaborating with the Royal Navy to develop this new adventure in aviation.

The technique of using aircraft from ships was very much in its infancy. To be honest, I hadn't done a single carrier landing when I took over as Chief Flying Instructor at Leuchars … and I was in charge, among other things, of the course training Naval and RAF pilots to land and take off from carriers.

First we made a mock-up of an aircraft carrier deck on the grass at Leuchars. On this, we practised being shot from a catapult to simulate being launched from a ship. After half a dozen 'take-offs' and 'landings', I flew down to Gosport with all the pupils on the course, to try our hand at the real thing, on the carrier HMS *Furious* in the Channel. She was one of the very early aircraft carriers: like a flat-iron, with no superstructure at all.

I was flying an Avro, just about the slowest of the aircraft for carrier landings. Normally, with a carrier landing doing about 20 knots, there would be a wind of about 30 miles per hour over the deck. The Avro landed at about 50 miles per hour, so the other two factors virtually reduced its speed to nothing.

You drew up in roughly 30 yards. This was just as well because the Avro had no brakes and the carrier had no arrester-wire, which later became vital.

Just to make sure you didn't overshoot, there would be a waiting deck party of sailors. As you touched down, they rushed forward, ran alongside the aircraft, then grabbed the wings and tail to bring you to a halt.

The greatest danger was side-slip, so they had to make sure you stayed in the middle of the deck. But should you go off at an angle while landing, wire nets were strung out along the side of the ship to prevent you going overboard. They were pretty effective, but not foolproof.

Some aircraft found the gaps in between, hit them and toppled over the top, or cleared them altogether, like a steeplechaser. Of course they fell into the drink, but the crew were usually picked up by one of the escorting destroyers.

With a good controller, it could be a pretty slick operation. The key to success was the lift in which we lowered the newly landed aircraft to the hangars below. The controller had to wait until it was back in position before giving the next plane the all-clear to land. Otherwise, if the lift stuck, the poor pilot was likely to find himself disappearing through a hole in the deck.

If conditions were favourable, you could land a squadron of about 10 aircraft in just about 30 minutes.

I'm happy to say that I had little difficulty getting the hang of it in practice, and it was just as easy when it came to the real thing. I was much happier about my appointment after tackling actual carrier landings myself. There were two distinct types of pupils under my wing. The first learnt to fly from scratch, then went on to carrier operation. He could be at Leuchars for about a year. The other was a trained pilot, converting to carriers. His course lasted only a matter of weeks.

There was only one serious accident and that was at Leuchars. One of the IIIF planes was being launched from the station catapult, to simulate a deck take-off. Everything went beautifully, except that the observer must have had his hand on the parachute release on his chest when the cordite charge fired the catapult. The jerk caused his hand to pull the rip-cord and the parachute shot into the air. As he was in the open cockpit, the slip-stream dragged him from the plane. He dropped onto the ground and was killed. The pilot didn't even realise he'd gone at first. It was only when he thought the aircraft felt a bit lighter that he looked round … and had the shock of his life to see an empty seat. But he still didn't know what had happened. He landed in a mixture of confusion and disbelief. He managed to do a shaky landing, but when he alighted from his cockpit he was distraught to learn what had happened to his observer, who was also his friend.

After two years at RAF Leuchars, in 1934, Brian was posted to the aircraft carrier HMS *Eagle*, as wing commander and senior RAF officer on the ship. Following a complete refit in the Far East, HMS *Eagle* returned to home waters where her personnel comprised 753 sailors and 253 airmen, together with their 9 Hawker Osprey Fighters and 12 Fairey III5s. In Brian's own written words:

This was a different way of life altogether, because it involved living in the ship, instead of dropping in occasionally as I used to do. I became more of a sailor than an airman and didn't take too kindly to my new way of life. Pitching and tossing around at sea wasn't my idea of RAF life.

My first stint of duty was a three-month spring cruise in the Mediterranean flying Fairey III/Fs and Nimrods. We put in a lot of hard work on aerial reconnaissance, air-to-ship gunnery and torpedo attacks. The torpedoes were real, but without warheads of course, and set to run underneath the ship.

We'd attack the *Eagle*, fire our torpedoes and watch the trail to make sure it passed under the ship, signifying a 'hit'. A destroyer then collected the spent 'dummy' torpedoes because they were too expensive to waste.

By this time, there had been new developments in carrier operation for aircraft. For a start, the carrier had arrester wires – just as well because the planes were landing at 60–70 miles per hour. If you missed the wire, you simply opened up the engine and took off again and made another circuit, hoping for better luck next time.

Another advantage of the arrester gear was that when the hook caught the wire, the aircraft was automatically centred on the deck.

Our big problems were rough weather and fog. Without any electronic aids, of any kind, it was no joke trying to put down an aircraft on the carrier deck, rising and falling through fifteen to twenty feet, in swollen seas and a pitching and rolling ship as well.

The difficulty in fog was finding the ship. Remember, we had no radio communications and relied solely on the skill of our observer, usually a navy man.

It was imperative that, before taking off, we knew the carrier's expected position at any given time in case of the weather closing in. About the only aid that the carrier carried was rockets to mark its position.

We also had to do night landings on an illuminated deck. This was also tricky. The deck had rows of little lights and there was a system of coloured traffic lights on the stern to help you judge height. As you came in, if you were lined up with the green, you were on the right glide path. If you went into red or yellow, you were either too high or too low.

After a long spell in the Far East, at the China station in the sweltering heat, the *Eagle* sailed west, arriving in the Mediterranean in February 1935, then a few months later she sailed for her home waters in Devonport, parting ways with Brian, who in May 1935 was transferred to another aircraft carrier, HMS *Courageous*, a deck-landing training carrier. She had capacity to carry forty-eight aircraft and was focusing on reconnaissance as well as anti-ship attack missions.

Brian picks up the story:

When the Italian invasion of Abyssinia came in 1936, I was in the carrier *Courageous*. We were posted to Alexandria, in case Britain should become involved. That was some show. We were tied up at the same quay as the Italian ships and watched them loading guns and ammunition to take down to the Suez Canal.

I had never seen so many warships in Alexandria. The harbour was packed. The British carrier HMS *Glorious* was also there. It was a tremendous naval force.

We sat about in Alexandria, doing exercises on the ship and getting up readiness in case we were needed. One day, a sleek, fast Italian ship from Brindisi came into the harbour. As I said, the congestion was terrible. There was hardly room to swing a cat.

In the middle of the harbour, the battleship *Royal Oak* was being ammunitioned, with the barges full of explosives alongside her. Well, the Italian dropped anchor close to the *Royal Oak* and swung round … then suddenly, for no apparent reason, it burst into flames. It was an incredible sight – and a great threat to the munitions on the *Royal Oak* and for all of us.

There was immediate panic across the whole harbour, lest the fire should spread, or the explosives trigger the whole show to go up and engulf all the ships. *Courageous* was only about 100 yards from the blaze. It sounds a lot, but it wasn't far enough when the ships were so tightly packed and the flames could soon leap through all of us.

The Italian ship blazed from stem to stern for three days. Fortunately, the flames were contained within the Italian ship. Even better, they got all of her crew off safely.

I was on board the *Courageous* for another 18 months of sea duty, up and down the North Sea, between Portsmouth and Scapa Flow.

We couldn't help wondering what on earth we were doing during this spell of messing about. It seemed obvious to us all that if war broke out, the carriers would be sunk at once. They were prime targets and far too vulnerable, operating in confined waters like the North Sea. Even the protecting screen of aircraft and destroyers wouldn't stop U-boats and enemy planes from reaching us. If carriers were to have any tactical value in war they had to be in the wide open spaces of the Pacific or Atlantic, where they were difficult to find.

I was glad to be posted to Abbotsinch as Station Commander before the war started, because I was certain the carriers would be the first to go. This turned out to be the case.

Chapter Ten

1937–1942

The Battle of Britain

As a new RAF station near Paisley in the West of Scotland (now
the site of Glasgow Airport), Abbotsinch had been designed,
built and equipped initially as a base for the development of
RAF-led 'Combined Operations in support of landing-craft training
and Radar calibration'.

Radar was a new science, only developed three years before – and quite
by accident. Its Scottish inventor, by the magnificent name of Robert
Alexander Watson-Watt, was trying to find a way to detect and avoid
the approach of thunder storms, but instead found he was developing
something much more important in a time of growing conflict.

Watson-Watt enthusiastically demonstrated his new discovery to
the Air Ministry at Daventry in an RAF Heyford bomber. They were
so impressed by this new defence tool that they immediately put in
orders. By the time war broke out in Europe in September 1939, a
whole network of radar installations were in place along the Channel
and the length of the eastern British coast. Brian was excited at this new
technology and had immediately invited Watson-Watt to Abbotsinch
to learn all he could about how radar worked and how to use it, both
on the ground and in the air, ready to train his pilots

Meanwhile, 516 Squadron flew in and prepared to get to work. Their
main brief was to provide air support for the combined operations,
training exercises and amphibious landings, involving all three services
– army, navy and air force.

Indeed, as soon as the navy and army troops arrived and approached
the beaches in their landing craft, the RAF pilots laid down a curtain

of smoke and strafing with blanks to simulate the conditions landing parties might encounter in a battle. This was an effective though not very welcoming introduction to their new base. In addition to their combined exercises, the squadron was tasked with calibrating the radar of the newly commissioned Fighter Direction Tenders, which were designed to provide forward radar as cover during troop landings. No doubt, Brian was quick to share his knowledge and train his men.

These and other plans were in preparation for a likely war. Indeed, there were already rumours about Germany's growing fleet of submarines, or U-boats as they were called, and their exercises in the Mediterranean. All three services understood the importance of being fully prepared ahead of time. For Brian, with his training background, he might have expected that his next posting would be in that role, but not this time – or was it?

After a year at Abbotsinch, Brian was sent to the other end of Britain to be station commander of RAF Gosport in Hampshire, with the rank of group captain. As well as a port and an RAF station with radar, Gosport benefited from the involvement of the combined services, with the valuable addition of civilian scientists and engineers. This multi-talented group also included the RAF Torpedo Development Unit. The civilians had their own wooden huts for offices and workshops, where they maintained and tested the various torpedo-launching aircraft, taking off and landing on the old parade ground.

As soon as he was settled in at Gosport, Brian started work on setting up a torpedo training course. This was an element that he was very interested in, so he was soon absorbed in all aspects of torpedoes and their use. This training began with daily practice in low flying over the sea, followed by learning to drop dummy torpedoes. These had to be dropped alongside a specific boat in Spithead waters, from where they were photographed. Brian showed each trainee pilot his photos, then discussed and analysed any errors with him. Similarly, any good drop was identified with the pilot and used as an exemplar. Either way, this training technique was very effective. Next would come mock attacks with 'collision-heads' against any of the obsolete naval vessels provided for this purpose. At that time, several different aircraft were used at Gosport for dropping torpedoes, but perhaps the most popular were the Hawker Horsley and the Fairey Swordfish.

To the south of Gosport was Stokes Bay Pier, which had been purchased by the Admiralty, together with the old railway line south from Gosport Road Station, for use as 'The Stokes Bay Torpedo Development Unit', under Brian's command. With a few adjustments – converting it to a narrow-gauge railway and making smaller trucks instead of the carriages – the railway was perfect for the conveyance and recovery of torpedoes, just as the pier at Stokes Bay was perfect for testing torpedoes.

Part of this involved descending the rope ladder that hung from under the pier's deck and jumping into the boat as it went by to observe each torpedo's performance, or collecting them afterwards.

The main purpose of all this was to observe the performance, accuracy and range of torpedoes when dropped from an aircraft (without explosives at this stage).

Being a man of action, Brian had dropped dummy torpedoes himself. This was to help him discover any potential problems, so that he could use his own experience, together with the speed and angle analysis of the scientists and engineers, in order to feed back their joint findings and advice to each pilot before his next drop.

In 1938, Brian was posted from Gosport back to RAF Leuchars as station commander. Before this, however, he and Jaimsie were invited to a Royal Garden Party on Monday, 18 July 1938 at Buckingham Palace. Perhaps they took the opportunity after that to visit Brian's sister and mother in Hertford, before their return to Scotland.

At Leuchars, Brian was based much closer to Jaimsie and their two daughters and he had always enjoyed training young recruits. However, now that a war seemed almost certain to break out within the coming year or two, Brian updated his training syllabus to make sure that he could impart all the new developments in flight and related matters, including torpedoes and radar.

At this stage, Britain was closely watching developments and setting the scene, as reflected by the national press, who still had their coverage of domestic events and all their usual features. In early summer 1939, the international peace was shattered and the RAF reconvened some of their specialist groups. Within days of the outbreak of war, Brian was promoted to Air Commodore. Then he was appointed Air Officer Commanding No. 15 Group. This was an RAF Operations and Training Group, coming under Flying Training Command and based in the south-east of England. This meant that Brian was now a director of flight training, to ensure the delivery of swift but comprehensive training to new recruits and more advanced training on aspects such as torpedoes and U-boats to more experienced pilots.

In this role, Brian travelled around Britain, inspecting and supporting RAF trainers. He clearly timed one of his visits to give him a few, very welcome, hours off so that he could take part in one of his favourite events – the opening of Tay Spring Angling in the Glendelvine water of the Tay. The local newspaper featured a photo of him proudly holding one end of a pole, with two other anglers holding up the other end, displaying Brian's catch of 'Five fine salmon'. Two of the salmon weighed 21.5 pounds each and the others almost as much. But that would be the last fishing trip for Brian for quite some time as German planes were now dropping bombs across the south-east of England. At first, this was sporadic, prompting the tongue-in-cheek tone of a small scrap of newsprint in Winifred's album:

To the Editor of *The Times*
Dear Sir, would you kindly thank Hitler for a wonderful rabbit dinner,
    Which took three 1,000-pound bombs and an oil-bomb to kill.
                    Yours truly,
        [signed by three Canadian soldiers]

As soon as he returned to his post, Brian found himself in the midst of a new and aggressive aerial campaign, waged by Germany – the Battle of Britain. On 15 June 1940, the British news headlines proclaimed that Hitler's tanks had triumphantly driven into Paris and up the Champs-Élysées. Churchill's response was, 'We shall not flag or fail.' With Europe capitulating to the Nazis, Hitler had drawn up a plan to bring Britain to its knees, ready to capitulate along with the rest of Europe – or so he thought. His intention was to send his air force, the Luftwaffe, to blitz London and rain bombs on shipping convoys, factories, airfields and ports.

Now, as a member of Coastal Command, Brian got straight down to planning counter-strategies and operations along the south coast. It was a fierce and relentless battle every day and night – the first purely aerial warfare between nations. However, as the Archives of Ontario explain, the British radar capability now came into its own:

> It was Radar more than the pluck of the dashing RAF pilots that tipped the scales in England's favour in the Battle of Britain.
>
> Hitler's strategic onslaught, meant to clear the skies over the Channel and south-eastern England, preparatory to an invasion of the British Isles, might have succeeded if not for Radar. The RAF was outnumbered by the Luftwaffe, and Radar saved the already stretched Fighter Command from having to maintain constant air surveillance.
>
> With Radar providing an early warning system, well-rested RAF pilots could be scrambled and ready to meet the incoming enemy formations in a matter of minutes.

Finally, three months later, Hitler had to stand down most of what was left of his prized Luftwaffe from his major objective. The battle peaked in September, with the blitz on London lasting a little longer, at great loss of life and destruction, but Hitler's assumption that the British would become demoralised and give in was very wide of the mark. He reluctantly withdrew and instead turned his attention towards Russia.

During the relative lull following Britain's first defeat of Hitler, Brian had some catching up to do – time for his family, for sports and for his old school. He also made time to fly himself to Ilkley to inspect a newly formed ATC cadet corps. He had no connection with that area, but he was always keen to support the cadets, wherever they were based. The newspaper photo showed the young lads standing tall and proud as Brian gave the salute.

One May evening in 1941, when Brian was enjoying a rare few days' leave in Scotland with his family, a bizarre situation was unfolding in the air. An unknown aircraft had been sighted crossing the North Sea, approaching the coast near Newcastle. The plane zigzagged at a decreasing altitude, as if lost. The mystery aircraft was given the code name of 'Raid 42'. The information was passed along and the plane's presence tracked. The pilot now flew overland towards Scotland. An Observer Corps volunteer on duty spotted the wayward aircraft and identified it as a Messerschmitt Bf 110. Flying at 50 feet and very low on fuel, the pilot parachuted out, landing near a farmhouse. The farmer and his son helped him out of his crashed aircraft and took him, limping with a possible broken ankle, into the house with his belongings. Curiously, these mainly consisted of homeopathic remedies and nearly thirty prescribed medications. The pilot introduced himself as Rudolf Hess and surrendered his pistol to the farmer. He then asked to be taken to Dungavel House to negotiate a truce with the Duke of Hamilton.

By now, it was almost dark. Both the army and the RAF had been alerted. The first to arrive were a corporal and a private of the Royal Signals, together with a guard. In the meantime, as the most senior RAF officer in Scotland that night, Brian received a phone call from HQ to go and take custody of Hess and find a secure place for him overnight.

Brian drove straight over to the farm, arriving late evening. On arrival, he had a brief chat with everyone, including Rudolf Hess himself, to establish what this bizarre situation was all about. Fortunately, Hess spoke good English and explained his mission was to make peace with Britain. Finally, Brian understood that Hess had not told Hitler and had embarked on this peace mission on his own initiative. He seemed hell-bent on independently ending the war between Britain and Germany by presenting himself at the residence of a certain Scottish Laird – the Duke of Hamilton.

Brian decided not to get involved with this unlikely plan, so when it was time to leave and Hess demanded to be taken straight to Dungavel House, he was resolute that it would not be possible at this time of night. With a promise from Brian to call the Duke of Hamilton in the morning, Hess acquiesced. Brian now detailed the Royal Signals contingent to take Hess under guard to the police station at Griffnock, then on to Maryhill Barracks, ready for a futile meeting with the Duke of Hamilton, as Brian had promised.

In November 1941, Brian was given a new and very different posting when he became Air Commodore and Commander of RAF Reykjavik, in defence of the North Atlantic. The British invasion of Iceland had been not only traumatic for its people, but also illegal and in normal times reprehensible. But this was a war. Iceland was a neutral country, with its own government and the loan of the King of Denmark, if ever they need a titular head of state. However, the Nazis had invaded and taken over Denmark and if that wasn't tricky enough, Germany was Iceland's biggest customer for its fish exports. Worse still for the Allies was the German diplomatic presence in Iceland. Britain offered assistance and protection to Iceland's government, but they had refused both, insisting on maintaining their neutrality, although that would not be on the table until the war was won. The main justification Britain had given to Iceland for the invasion was protection for them and other Atlantic nations, which was genuine. However, the most pressing purpose for the Allies was to prevent Germany taking over Iceland. Brian realised that he would need all his reserves of diplomacy in this new role.

The Canadian Army sent troops to help the British Army with their infrastructure programmes, the most pressing of which was to build an airport. Some American troops also came to help, although the USA would not join the war for a long time yet. This was the situation Brian found when he flew over to the newly completed RAF Reykjavik air

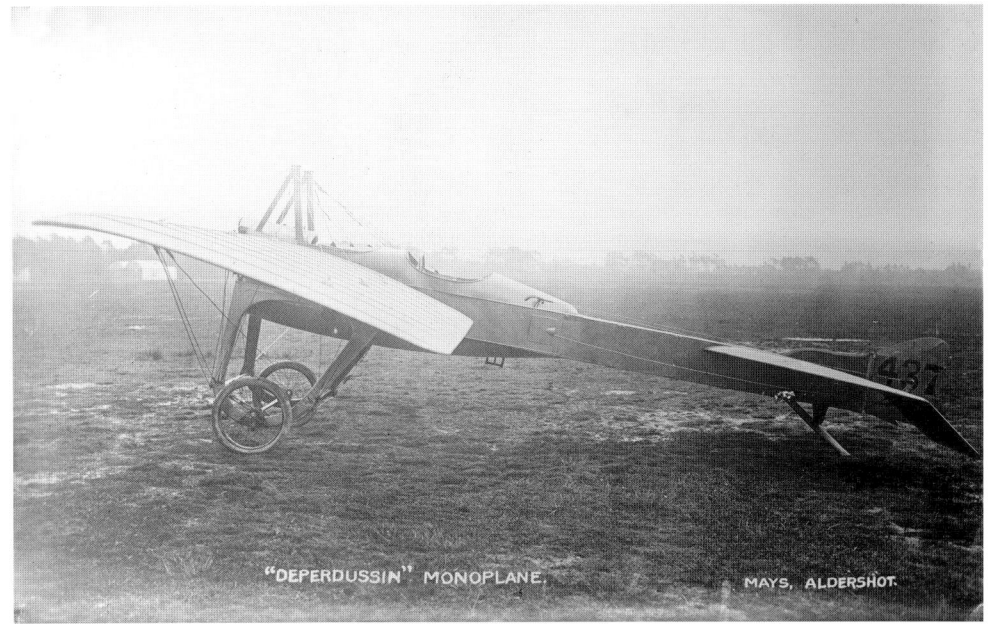

A Deperdussin TT, an early RFC aircraft. (*Wikimedia Commons*)

The Deperdussin crash debris. Brian is on the right, next to the girl wearing a white blouse. (*Stevenage Museum*)

Brian, 2nd Lieutenant, 15th Rifle
Brigade, pictured in the summer of 1915.
(*Family archives*)

Brian on leave, September 1915, before joining the RFC. (*Family archives*)

Montrose airfield, Broomfield, 1914. A BE2 biplane can be seen in the air, with two MF7s in the foreground. (*Montrose Heritage Centre*)

A Martinsyde biplane, one of the aircraft in which Brian learned to fly. (*Wikimedia Commons*)

A Maurice Farman Longhorn, in which Brian gained his pilot's licence, in record time. (*Wikimedia Commons*)

The page in Brian's pilot's log book, on which he had written his destination as 'French Aerodrome'. (*Family archives*)

A Bristol Scout, the aircraft Brian flew over the Channel to war. (*Wikimedia Commons*)

The Bristol Fighter BE2c, Brian's first combat aircraft. (*Wikimedia Commons*)

Brian in his pilot's boots at the front of his Airco DH.9A bomber, surrounded by his maintenance team. (*Family archives*)

A Bristol Fighter, like Brian's, in the midst of a series of dogfights with the Germans. (*Wikimedia Commons and family archives*)

Autumn 1917: a new generation of fighter aircraft included the SPAD S.XIII. (*Wikimedia Commons*)

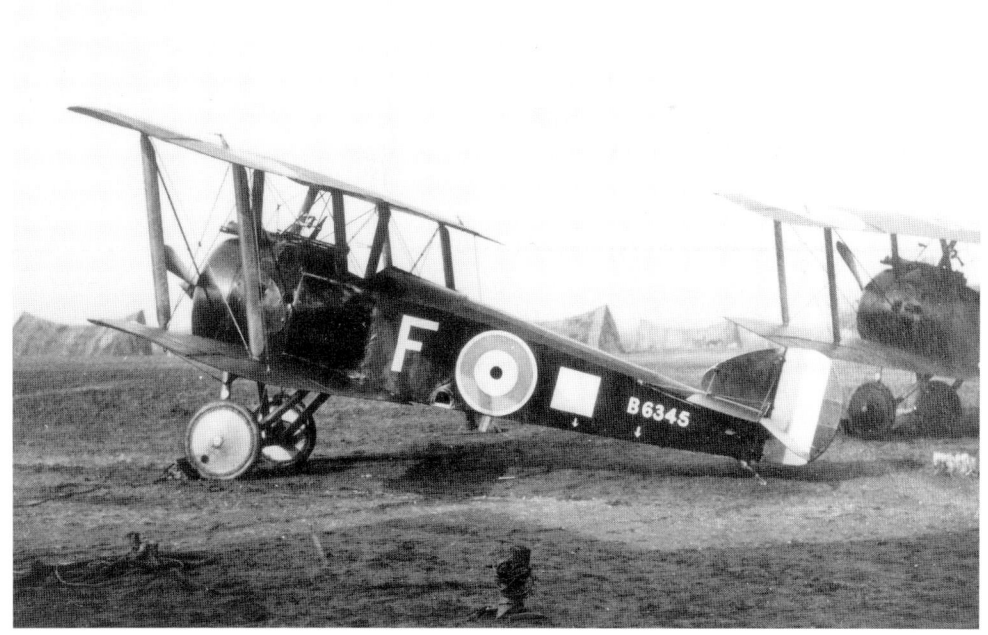

The Sopwith Camel. (*Norman Franks*)

The SE5a. (*Wikimedia Commons*)

A souvenir brought back home by Brian to the family home in Hertford, 1917. (*Family archives*)

In December 1917, Brian and other pilots dropped cheeky Christmas cards along the German trenches. (*Family archives*)

In 1918, the French president awarded Brian the *Croix de Guerre* with Palm. (*Family archives*)

Anthony Fokker in the cockpit of a Fokker F.I prototype in 1917, with the Red Baron standing third from the right. (*Dutch Ministry of Defence via Wikimedia Commons*)

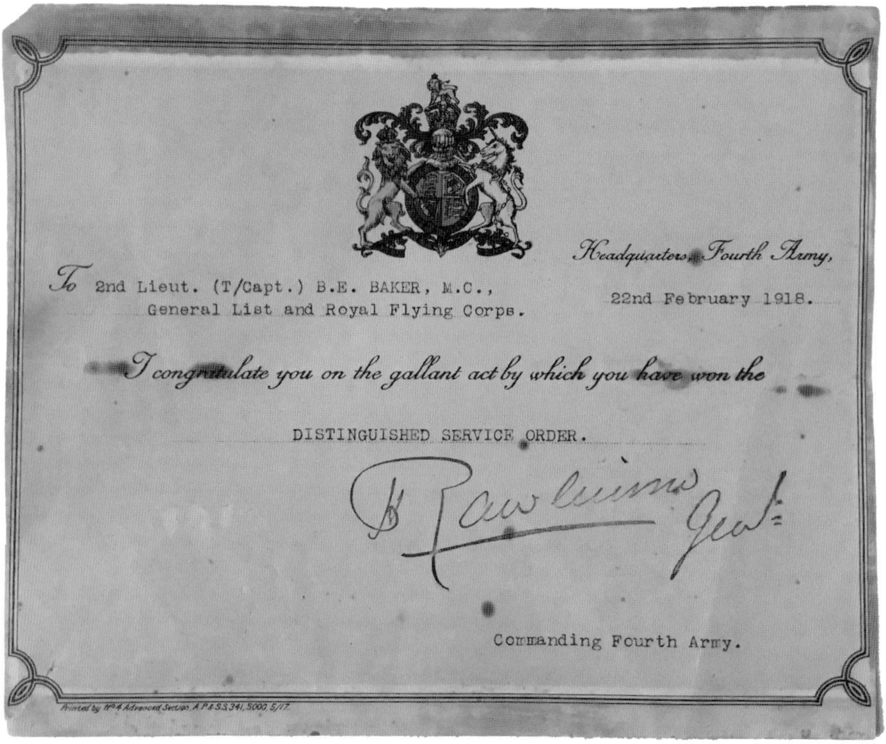

Headquarters, Fourth Army,

22nd February 1918.

To 2nd Lieut. (T/Capt.) B.E. BAKER, M.C.,
General List and Royal Flying Corps.

I congratulate you on the gallant act by which you have won the

DISTINGUISHED SERVICE ORDER.

H Rawlinson Genl.

Commanding Fourth Army.

Printed by No 4 Advanced Section. A.P.&S.S. 341. 5000. 5/17.

Brian was awarded the MC and DSO. (*Family archives*)

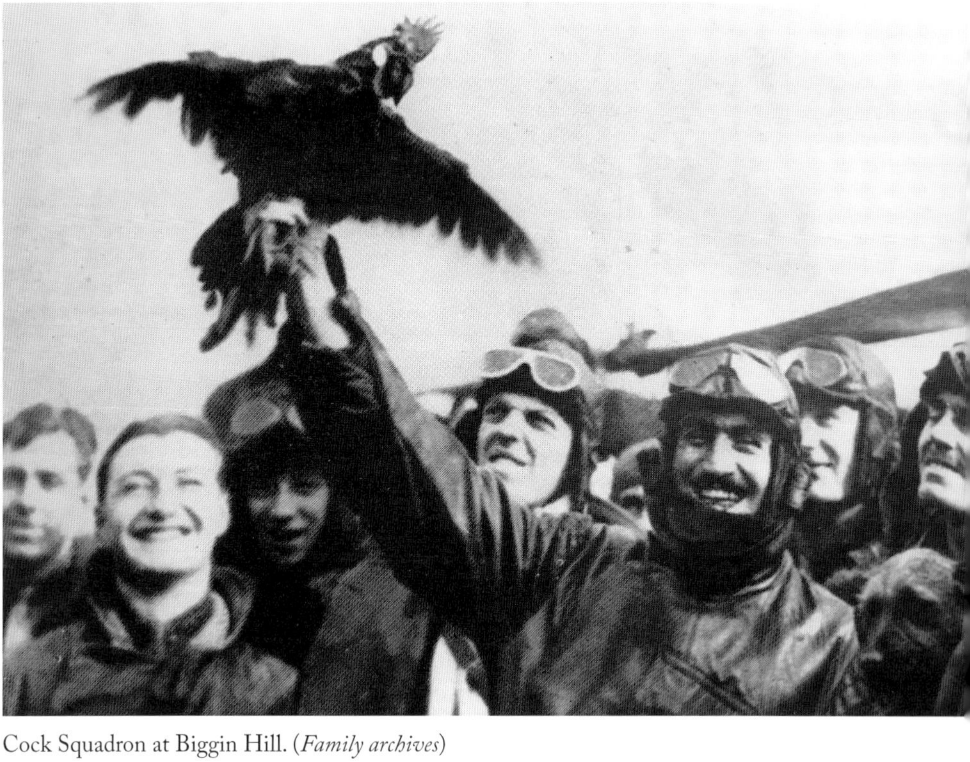

Cock Squadron at Biggin Hill. (*Family archives*)

In 1921, Brian's appointment as Chief Flying Instructor at RAF Abu Sueir in Cairo. (*Family archives/unnamed newspaper cutting*)

Shepheard's Hotel in Cairo, where Brian stayed and met Jaimsie, his wife-to-be. (*Wikimedia Commons*)

Brian and friends visited the newly discovered tomb of Tutankhamun. (*Wikimedia Commons*)

Back in command at RAF Biggin Hill, Brian organised the first public demonstration of telephony at the adjoining Wireless Unit. (*Family archives*)

Brian was in charge of experimentation at the Royal Aircraft Establishment. (*Wikimedia Commons*)

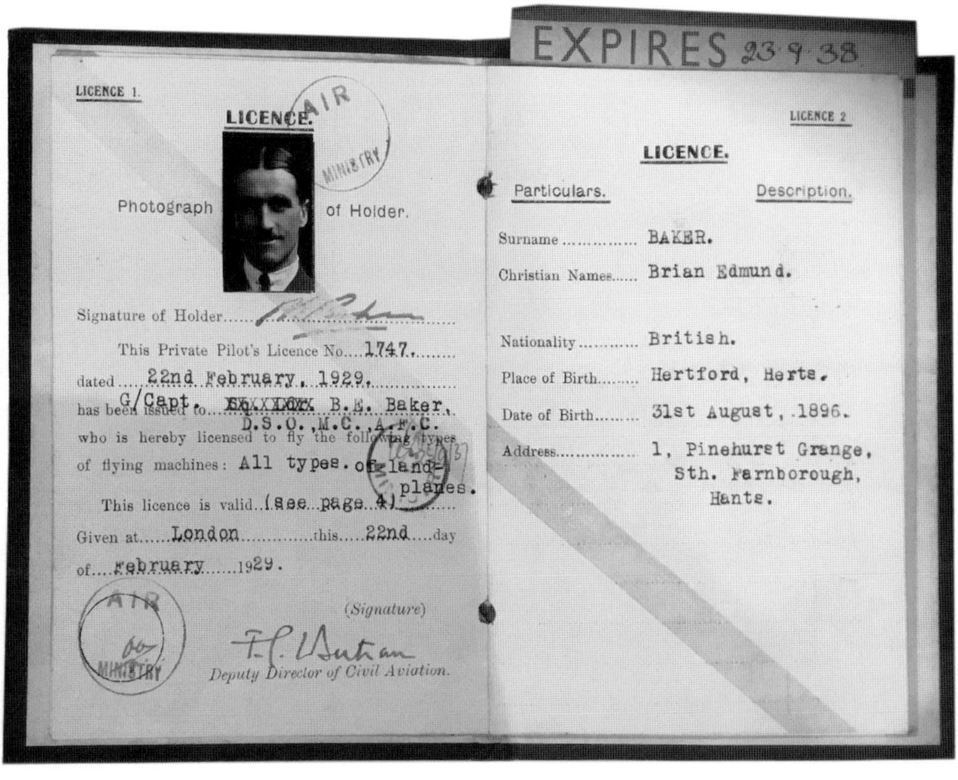

EXPIRES 23·9·38

LICENCE 1.

LICENCE.

AIR MINISTRY

Photograph of Holder.

Signature of Holder......

This Private Pilot's Licence No....1747....

dated .....22nd February, 1929.

has been issued to.....G/Capt. BRIXTON B.E. Baker.
D.S.O. M.C. A.F.C.

who is hereby licensed to fly the following types
of flying machines: All types of land
planes.

This licence is valid..(see page 4).

Given at......London......this....22nd....day

of....February......1929.

(Signature)

F.C. Shutan

Deputy Director of Civil Aviation.

LICENCE 2.

LICENCE.

Particulars.   Description.

Surname .... BAKER.

Christian Names...... Brian Edmund.

Nationality .... British.

Place of Birth........ Hertford, Herts.

Date of Birth........ 31st August, 1896.

Address............... 1, Pinehurst Grange,
                        Sth. Farnborough,
                        Hants.

Brian passed his private pilot's licence in 1929. (*Family archives*)

Tasked with working out how to design and use the workings of aircraft carriers, Brian and his students experimented with a giant catapult. (*Family archives*)

In 1935, Brian (bottom right) watches trials as an aircraft with folded wings goes down in the lift of HMS *Courageous*. (*Family archives*)

Official war artist William Rothenstein's sketch of Brian exhausted at the end of a sortie. (*Family archives*)

n 1939, Brian's favourite aircraft carrier, HMS *Courageous*, was struck by a U-boat; it caught fire and sank, ith the loss of 518 men and its full complement of Fairey Swordfish aircraft. (*Public domain*)

Brian (second from left), appointed C-in-C of Coastal Command, with his team, planning Operation Cork as a vital part of D-Day. (*Coastal Command*)

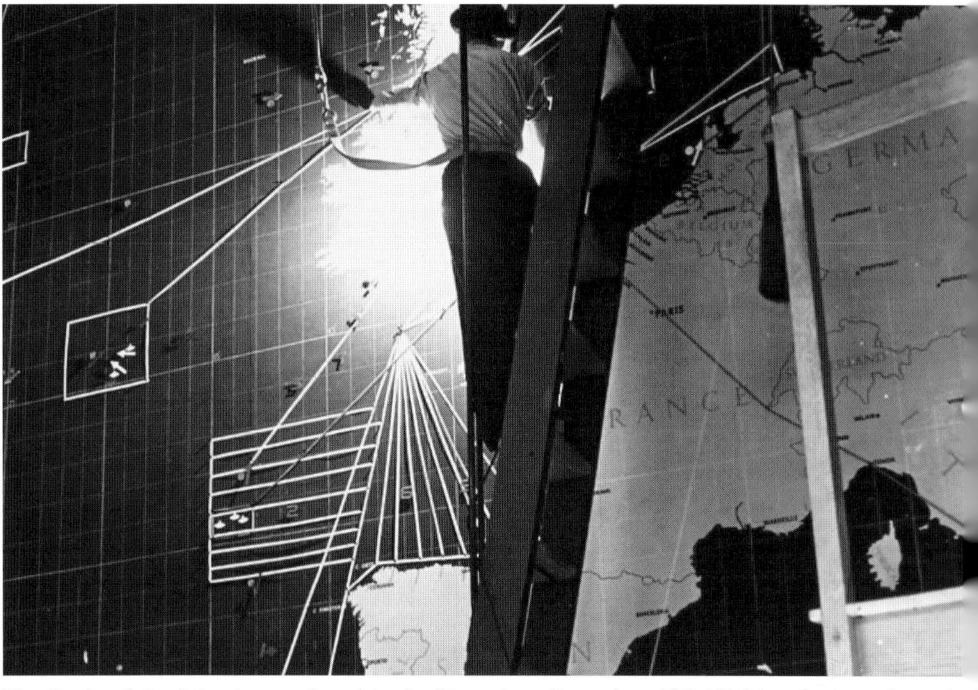

The Battle of the Atlantic was plotted in the Operations Room by a WAAF. Here she is tracking the position of a convoy. The white lines are aircraft sweeps and the small white marks denote U-boat sightings. (*Coastal Command*)

The crew of 'H for Harry', a very long-range Liberator, don their Mae Wests, put their kit on board and set out on an eighteen-hour U-boat hunt. (*IWM*)

B-17 Flying Fortresses over Schweinfurt, Germany, 17 August 1943. (*Public domain*)

This 'lovely but sinister' Beaufort is armed with its torpedo and ready for take-off. (*Air Force Museum of New Zealand/public domain*)

H.M. KING FAROUK I

Levee at Abdine Palace on 6th May, 1945

AirVice Marshal Sir Brian E. Baker;
KBE., CB., DSO., MC., ⬛C. ⋅
will be presented by

Air Officer Commanding-in-Chief:

Instructions are on the reverse of this Card

King Farouk invited Brian to a weekend party at his Abdeen Palace. (*Family archives*)

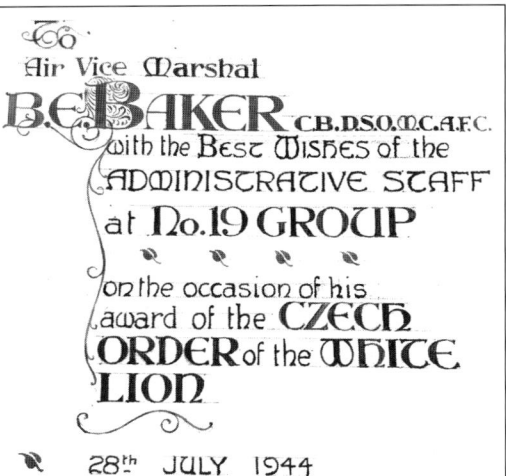

To
Air Vice Marshal
B.E.BAKER C.B.,D.S.O.,M.C.,A.F.C.
with the Best Wishes of the
ADMINISTRATIVE STAFF
at No.19 GROUP
on the occasion of his
award of the CZECH
ORDER of the WHITE
LION

28th JULY 1944

Brian was awarded the Czech Order of the White Lion, their highest order, on 28 July 1944. (*Family archives*)

As C-in-C of Transport Command, Brian now had the task of organising the British and Colonial efforts to provide for the besieged Berliners with essential supplies and services. (*Family archives*)

This aircraft is about to drop a large bag of candy bars for Berlin's children. (*USAF/public domain*)

Brian and Jaimsie
at a function.
(*Family archives*)

Official RAF portrait of Brian.

Part of Brian's role at Transport Command was to accompany royals to and from their train, plane or ship. On this occasion, Brian escorts Princess Elizabeth to see Prince Philip off from London Airport to Malta. (*Family archives/unnamed newspaper cutting*)

As well as helping the veterans in need, Brian organised reunions for both RFC and RAF veterans. This was a combined reunion with travel laid on free. (*Family archives/unnamed newspaper cutting*)

For royal visits to Scotland, Brian was often invited. Here he accompanies Queen Elizabeth II as she inspects RAF veterans on parade at Holyrood House. (*Family archives/unnamed newspaper cutting*)

Brian with Air Vice-Marshal T.L. Kennedy, Air Officer, Scotland, after retiring with twenty-one years' service as Scottish Branch Secretary of the Royal Air Force Benevolent Fund, August 1978.

Brian's wife, Jaimsie.

After Brian's death, a plaque was placed in Hertford's parish church in the presence of Air Marshal Sir Edward Chilton KBE, CB (second from left), his daughter Margot (right) and his devoted sister Winifred (second from right), without whom this book could not have been written.

Jaimsie and Brian's joint grave at St Andrews, Brian's favourite place. (*Family archives*) base, along with a sizeable contingent of pilots and ground crews. With the RAF's arrival adding to the army and navy, the combined military forces outnumbered the island's male population, which caused great concerns amongst islanders and inevitably led to many 'unofficial' liaisons, with high numbers of mixed nationality births down the line.

This furore was one of the first hostile problems that confronted Brian. As commander, he had to meet with his naval and army counterparts to prepare a joint meeting with the island's parliament, not to mention some very angry citizens. Although there remained a number of other problems, by now the islanders could see for themselves how the army turned their dirt tracks into roads, whilst the navy patrolled the seas and the RAF kept watch of the skies, to keep them all safe from Nazi bombers and U-boat attacks. Indeed, Brian keenly joined the effort whenever he could, reconnoitring from the air, chasing any Luftwaffe aircraft away.

Communications with Denmark and Germany had been closed since the invasion, but engineers provided a network of phone lines and electricity across the island, which were much appreciated by the inhabitants, as were their increased incomes from the newly opened bars and restaurants for the forces in Reykjavik.

From first setting foot in Iceland, Brian was keen to explore, and what he found pleased him greatly, as he told his sister in his first letter home to her from his new posting:

Hotel Borg
Reykjavik
31st March 1942

My dearest Win,
I hope you will all have a very happy Easter and plenty of eggs with some nice weather in which to consume them.

I had a very interesting trip of about 50 miles along the south coast a few days ago. The road was beautiful and there was a good wind on. The surface was solid ice, but with big tyres it did not seem to make any difference and we got along very well.

There was nothing but ice and snow and lava rock as far as the eye could see, until we suddenly came upon a ski-jump and a pretty little chalet, owned by the ski-club. Just around the corner, to my amusement, was a large jet of steam coming out of the middle of the solid ice. I then discovered that the heating of the chalet is all obtained from this hot spring.

Further on we came to a small village, and any number of greenhouses with steam shooting out all over the place. Tomatoes etc., growing in the greenhouses, all heated by the steam.

I saw one place to the side of a huge piece of rock, where the lava looked as if it had just bubbled out and cooled off. I'm sure the whole place will blow up one of these days, especially if the steam jets happen to get blocked up.

This pub is quite comfortable, but the heat is awful. I have all the windows open in every room to keep it within bounds. Perhaps it is OK in the winter, but by heaven, it will have to be cold to bring the temperature low enough to be comfortable.

I had a look round the shops. There seems to be plenty of everything and, provided you go to the canteens and officers' shop, it's quite cheap.

I have got some fishing under way, but it will not really start until June. They think nothing of salmon up here and are not interested in trout. I asked for some trout fishing and they scornfully said, 'Oh you can get some little ones in certain rivers and lakes not far from here.' So I rather gingerly asked what they called little ones and he said, 'Oh, about three or four pounds.' I said (with my eyes popping out), 'Of course that is not very big, but twenty or thirty, which I'm informed is a normal evening's work, would be quite good form every now and again.'

My bungalow is situated in a most convenient place for walking out on a rock and catching a nice cod for lunch.

It is getting on here and it will be ready for habitation next month. I must start collecting furniture, etc.

Perhaps some lemons and an orange or two will be turning up for Richard in the near future. Don't expect them until you see them.

I expect to get a letter or two by air mail today. I hear it is on the way. Letters in and out of here are certainly dependent on the weather.

I have no more news that I can tell you, and plenty that I cannot.

love to you all,

Yours Affec.

Brian

Brian's mention of this bungalow suggests that, whilst all personnel had been housed in adequate temporary accommodation from their arrival at RAF Reykjavik, including the commander himself, construction was ongoing to improve living quarters and facilities, as well as recreational provisions for all ranks.

As ever, Brian carefully avoided telling Winifred about any of the dangers, but throughout the time that he was at RAF Reykjavik, there were attacks by both sea and air. On one occasion, a German fighter plane managed to fly straight into the island's airspace, but fortunately was taken down before it could drop any bombs. Nazi U-boats were the main and frequent danger, but all the navy and air force training and their vigilance in patrol, reconnaissance and anti-submarine duties successfully repelled or destroyed most of them. However, the records show that during the course of the Allies' occupation, over 200 Icelandic seamen were killed by Nazi U-boat attacks.

Brian knew that most of those deaths could have been avoided, if only Iceland had been given its own radar set.

Chapter Eleven

1942–1944
U-boats

It was just as well that Brian hadn't yet bought any furniture for his Icelandic bungalow, because everything was about to change. It was late June 1942 and, unexpectedly, after just four months in Iceland, he was called back to England by Coastal Command, who needed his expertise in reconnaissance, radar and torpedoes. He was appointed Air Officer Commanding No. 16 Group, Coastal Command.

Brian arrived back at HQ and the following day, 2 July 1942, was promoted to the rank of air vice-marshal. He received many congratulations, including from his old school, Haileybury, together with an invitation to attend a special fundraising event in 'Wings for Victory' week. He agreed to come along and bring a couple of friends.

The Haileyburian magazine described the week's Wings for Victory programme:

> The Cadets have taken part in many outside activities this term, in connection with the local 'Wings for Victory' Week.
>
> A flight of 50 cadets took part in the opening procession at Ware. On 29th May a Guard of Honour was mounted at the Village Hall, when Air Vice-Marshal Brian Baker attended an open-air Church Parade on the village green.

The events continued, culminating in 'a Ceremonial Parade and inspection on Terrace Field by former Haileyburians, Air Marshal J.C. Slessor, Air Vice-Marshal B.E. Baker and also Colonel Porter of the American 8th Army'. Whilst appreciating as role models the two former pupils, both now members of the British aeronautical hierarchy, it is not hard to imagine the boys' excitement in having a real American colonel taking an interest in these young English cadets.

Brian returned to Hertford many more times to help with fundraising or to accept donations at the Corn Exchange, the school and various other locations and events, or to read the Sunday lesson in church. Of course, on each occasion he also spent time with his mother and his sister Winifred, but only a day or two. He was needed back at his base on the south coast, defending southern England.

As commander, Brian planned reconnaissance and practice exercises for his squadrons, to keep everyone fit and ready for anything. Like all commanders, he had to keep a daily log of activities, weather conditions, events and welfare. Here are recounted four consecutive days of Brian's 1942 log entries (unaltered) to demonstrate the variety of experiences:

12 April

Officiating at 1015 hours. Pilots called for a roodoo [a briefing to senior officers]. They get off at 1127 hours, crossing the coast at Beachy Head. Proceed to Le Tréport and Ambleteuse and back to base. This was quite an uneventful operation. Formation flying in the afternoon was followed by debriefing at 1730 hours for a circus. The wing to act as target for Boston bombers in bombing. Has been carried out according to plan. Visibility was poor and no opposition was encountered. Squadron to act as cover patrol. The wing took off at 1800 hours, landing again at Abbeville. We are at readiness again from 1700 hours to 2310 hours for assault. Was cancelled.

13 April

Sgt. Forbes (Polish) was posted to no. 16 (Polish) Squadron today. At 1325 hours the Wing was called to briefing at the target near the marshalling yards at Boulogne with no. 8 Squadron to act as cover escort. Led by W/C (Wing Commander) B. Finucane. The Squadron took off at 1408 hours and made rendezvous at

Romney Bay with Boston bombers and North Weald Wing at 1430 hours. Returned at 1500 hours, the operation having been successfully completed.

14 April

Weather continues fair. F/sgt. (Flight Sergeant) D.P. Bushband (Can) Ostend to no.81 Squadron. Pilots were called for briefing at 1710 hours for a roadstead to Ostend, taking off at 1811 hours. The object was reached at 1800 hours. Three minesweepers were well-staffed. S.B. Moston, wishing to be quite fair so gave ground defences a share of ammunition. P/O (pilot officer) Shaw D.P.W. Landed at 1853 hours, having had to leave the Squadron and return to base as was suffering from spots in front of his eyes. Reputed to be flies on his windscreen. P/O A.E. Shackleton made a landing in the North Sea and was quickly picked up by an Air-Sea Rescue launch, his theories about flak having been shattered.

15 April

A dull, cloudy day. Sgt. Bloodhart posted to no. 81 Squadron today. Pilots called for a briefing at 1125 hours, the target being a hutted camp at Etaples. They took off at 1152 hours, reaching the target at 1236 hours and returned home at 1300 hours. Considerable flak was encountered in the target area, 3 of our aircraft having been damaged. F/Lt. J./ T./ Weller's (Can) machine received several hits and he made a fine landing on his return to base

Heavy loss today. Wing Commander B. Finucane's machine was landing on the sea about 10 miles west of the French Coast. His machine sank immediately and no further trace was seen of him. It was assumed that he must have been stunned on impact with the water. He will be greatly missed by no. 81 Squadron, with whom he was very popular. Spending a considerable time at our disposal, it is a very heavy blow to lose a brilliant fighter-pilot at the early age of 21.

In early 1943, Brian was transferred to Coastal Command Group No. 19, with responsibility for reconnaissance and defence. In a newspaper article he wrote, he explained the operations that his group of Coastal Command carried out in preparation for the big day – the secret plan, code named Operation Cork.

> While I was in command of no. 19 group at Plymouth, our task was to look after the South-Western approaches. The German U-boats were stationed in French ports, ready to strike at any time, so the RAF were on almost 24-hour watch to protect both naval and commercial shipping moored at Plymouth or passing through.
>
> In May 1943 the enemy adopted a policy of going out to the Atlantic, on the surface in groups of three and fighting it out with our anti-submarine aircraft. This rather suited our book. We bagged a lot of U-boats during that month.
>
> During all this anti-submarine war, we were very ably supported by the Royal Navy. The co-operation between us and the hunting groups of frigates and destroyers was 100%.

All this time, Brian was alternating between planning and flying, though rather less flying than he would have liked. On 3 March 1943, he and Jaimsie were invited to another Royal Garden Party at Buckingham Palace, an event that they always enjoyed. In mid-March, Brian received another familiar looking royal envelope. When he opened it, he found an invitation for him and Jaimsie to a royal party on 31 March, but it wasn't a garden party. Puzzled, Brian read on and discovered it was to attend an afternoon and evening party at Windsor Castle with King George VI and Queen Elizabeth. He unfolded the invitation, which was the programme for the party, phoned Jaimsie and read out the plan of activities:

> Musical selections by guest artist and other music from:
> Father McNally
> Private Deddo

The Scutlebut Five

Tankersley Amil on the organ
Holmberg and Lynch
Grand Lottery and games, with prizes
Presentation
Beer and Ice cream

In the evening were 'Movies', although there was only one film: *This Girl Friday*, starring Cary Grant and Rosalind Russell. It was a pleasant interlude from the war for Brian – but not for long.

Whenever possible, Brian was still in active service with his squadron. In May, they were out reconnoitring the western end of the Channel and the eastern Atlantic when he noticed that, instead of the usual lone U-boats they had previously seen and dispatched, this time the Germans seemed to have adopted a new approach – that of prowling and surfacing in the Atlantic in packs of three and fighting it out with the Allies' anti-submarine craft.

On 2 June that year, in the King's Birthday Honours, Brian, was awarded the Companion of the Order of the Bath (CB), followed one month later by his investiture at Buckingham Palace. He received several congratulatory telegrams, including one from Viscount Archibald Sinclair MP, from the Air Ministry (and a friend of Churchill): 'Please accept the heartiest congratulations from the whole staff of Commander-in-Chief on your CB.'

Soon after this spate of royal events, Brian was back at base when he received an emergency call that needed all hands on deck – literally. He called for all the help he could muster, directed and despatched ships and aircraft, piloting one himself. A newspaper described the events:

101 MEN RESCUED AT SEA AFTER RAIDS ON GERMANY

In the greatest air-sea rescue operation of the war, 101 British and American airmen have been rescued from the North Sea, within

50 hours and have been landed at different points in England. All of them had been forced down into the water through damage to their aircraft in the recent heavy day and night bombing raids on Germany.

On Sunday, after Bomber Command's massive assault on Hamburg, the United States Fortresses raided northern Germany and 19 of them were reported missing. The first S.O.S. from one of the Fortresses was received by a Coastal Command group's flying control officer in the late afternoon. As the attacks on industrial Germany continued on a vast scale – Bomber Command raided Essen on Sunday night and American Fortresses raided Hamburg in daylight on Monday. More reports of 'ditched' aircraft came in and rescue operations intensified.

High-speed launches, Walrus amphibians, lifeboats, trawlers and fishing smacks co-operated in the search for survivors, while the boats carried by Air-Sea Rescue were dropped, and frequently to aircrews nearly 200 miles apart, during the 50 hours ending 7.30 p.m. on Tuesday.

More than 200 aircraft from Bomber, Fighter and Coastal Commands, as well as aircraft of the U.S.A.A.F. participated in the operation, searching by day and night and guarding the dinghies and lifeboats, launches and Walruses from possible enemy interference.

Most of the rescues were made more than 100 miles from England, about half-way across the North Sea. A few airmen were seen by the Royal Observer Corps and rescued within sight of the English coast. But one crew of nine Americans were saved when an airborne lifeboat was dropped by parachute, approximately 200 miles from England and only 60 miles from enemy soil. At one period, more than 70 aircraft were in the air at once – Fortresses, Halifaxes, Stirlings and Beaufighters.

After it was reported on Monday that a Fortress had ditched 60 miles north of Borkum, an airborne lifeboat was dropped and nine Americans scrambled into it. They were given protection until darkness. On Tuesday their lifeboat was located again and additional supplies of oil and petrol were dropped.

Then it was reported that a foreign trawler had stopped the lifeboat and had taken them to enemy territory. A British Halifax aircraft flew over the trawler and persuaded it to turn about and proceed towards England. On the way across the sea it was escorted by the R.A.F.

Two German survivors of the aircraft shot down off the Yorkshire coast were found floating in a rubber dinghy by the Air-Sea-Rescue Service. When shown the newspaper report of Mussolini's fall, they laughed and refused to believe it.

In August 1943, Brian noted that the enemy had changed their U-boat tactics again: 'They obviously could not stand their heavy casualties.' They began to surface for the shortest possible periods during the night.

As a result of our co-operation with the hunting groups, we achieved intensive study and training to help us in directing surface craft by our aircraft who spotted the U-boat and communicated it their exact position, to achieve greater accuracy. This would prove to be of immense value at the start of the planned invasion, when the time came.

Although often heavily involved in the planning for Operation Cork, often with his team at Coastal Command HQ, Brian continued whenever he could to join his fellow pilots patrolling the seas of the North and the East Atlantic. On one occasion, they spotted and sank three U-boats, which was a considerable feat for such a low-key outing.

On another night, in September that same year, Brian had planned a longer, more ambitious and more complex patrol, which included the involvement of some of their USA and Canadian Air Forces friends, together with the support of the Royal Navy. The following morning, the Admiralty's announcement about this night made headlines in the national newspapers. An extract from one report gives a vivid account of the action:

A NOTABLE VICTORY

The destruction of seven German submarines in the Bay of Biscay was announced by the Admiralty last night. They were sunk in a series of combined actions by ships of the Royal Navy and British and American aircraft. Naval and Air patrols had been increased to counter the enemy's attempts to send groups of two or three submarines across the Bay in company.

In one encounter, described as a notable victory, three U-boats were sunk, two of them supply submarines.

The Admiralty statement says: 'In the first of the actions which resulted, a Sunderland aircraft and a Halifax of no. 19 group Coastal Command Air Vice-Marshal B.E. Baker, CB, DSO, MC sighted three U-boats proceeding at full-speed on the surface. As the aircraft manoeuvred into position for attack, they were met by a barrage of anti-aircraft fire from the enemy.

FOUR DIRECT HITS

Ultimately the U-boats became separated. In an attempt to disengage, they started to submerge. The Sunderland aircraft immediately flew low over one of the U-boats and carried out a skilful attack with depth charges.

One fell on the conning tower and three more rent the hull near the starboard bow. The U-boat sank within a few minutes, leaving more than twenty of its company in the sea.

Another single U-boat was attacked and mortally damaged by a Liberator aircraft of the US Army Air Force anti-submarine group (Colonel Howard Moore). As the enemy settled by the stern, it was dispatched finally by a British Halifax aircraft. Between 30 and 50 survivors were seen in the water.

A few days later, a Wellington aircraft severely damaged another solitary U-boat. About 20 members of the ship's company appeared on deck as a second Wellington aircraft. Within a few minutes, the enemy submarine blew up.

A third single U-boat proceeding on the surface was encountered by an American Liberator aircraft. Members of the ship's company attempted to man a gun, but they were swept from the deck by the force of the exploding depth charges as the aircraft made two runs over the target.

A British Liberator made a further attack on the U-boat, which then disappeared. Shortly afterwards it broke the surface again, keeled over and sank, leaving wreckage and ten men in the sea.

Two days later a combined action by sloops of the Royal Navy and by aircraft of Coastal Command and the United States Army Air Force won a notable victory. It cost the enemy three U-boats, including two of his scanty force of supply submarines. A second Wellington Aircraft joined in the attack.

And so the battle continued …

A necessary postscript to these Allied successes is that in all possible cases the Royal Navy picked up survivors and took them back to Plymouth for any necessary medical treatment. They became prisoners of war and were treated humanely for the duration.

The above account goes on to describe a number of successful follow-up actions as the war progressed, all of which helped to lay the ground for Operation Cork and the big day that would be make or break in its intention to liberate Europe from Hitler's Nazi tyranny.

Chapter Twelve

1944

Operation Cork

Both sides of the war were exploring new ideas and developing new instruments and weapons, as well as testing new tactics and modus operandi.

At the end of 1943, Hitler was keen to rush out the Germans' latest invention, or rather an improved version of something they had been working on since 1939 – glider bombs. Although not yet fully tested, the Nazi hierarchy decided to use their glider bombs as their best weapon in stopping and destroying the Allies' large, important and very strongly guarded convoy that was moving about 250 miles south-west of Cape St Vincent on its way towards Britain.

Brian's No. 19 Group of Coastal Command, augmented by USA and Canadian Air Forces, were protecting the convoy as it slowly moved towards its destination. One of many newspaper reports detailed the story about the tortuous four days' and four nights' journey of this convoy:

DOUBLE ASSAULT ON CONVOY
DEFEAT OF NEW TACTICS
GLIDER BOMBS FAIL

The decisive defeat of attacks by U-boats and aircraft with glider bombs on an Atlantic convoy was announced by the Admiralty and Air Ministry on Saturday, in the following combined statement:

Attempts by strong forces of U-boats and later long-range enemy aircraft armed with glider bombs, to launch a major attack on an important Atlantic convoy were decisively defeated by H.M. escort ships in cooperation with aircraft of Coastal Command and aircraft of the U.S. Navy operating with that Command.

During the long period of the convoy's passage, close escort was provided, day and night, by surface forces of the Royal Navy and Royal Canadian Navy and by Catalina and Ventura aircraft of the U.S. Navy, by Hudson and Fortress aircraft of Coastal Command, operating from Gibraltar and the Azores, and by Liberator and Sunderland aircraft of Coastal Command, operating from British bases.

In a series of engagements which continued intermittently for four days and three nights, H.M. ships destroyed at least one U-boat, probably a second and damaged several others. In addition to these successes, one U-boat was probably destroyed and another believed to have been damaged by Leigh Light aircraft of the Coastal Command.

A number of enemy aircraft were shot down into the sea and some others were so badly damaged that they were unlikely to reach their bases.

The glider bombs had ignominiously failed and all attempts by the enemy to launch a counter-attack were completely frustrated.

With dawn came the knowledge that the convoy, with its ring of protection, was within reach of its destination ... but it wasn't quite over. From their bases in France, several hostile bombers, including seven Focke-Wulfs, were sighted approaching the convoy, but now the Allies' fierce anti-aircraft fire did so much damage that the surviving Nazi aircraft withdrew, just three minutes after their arrival. Only two of the convoy's ships were damaged, and only slightly, so it was a very good result for the Allies.

Throughout 1943, as a key member of Coastal Command, Brian had worked with his handpicked team, progressing their plans. Brian was in charge of the dangerous south-western flank. Despite the Allies' magnificent efforts, reconnaissance indicated that Germany still had approximately forty U-boats in French ports, so it was going to be a mammoth task to fulfil their goal of ensuring not one German submarine could access the Channel to threaten Britain's planned invasion of Europe, which was aptly designated Operation Cork.

Working with Brian on their secret plans were four trusted team members: Air Commodore N.H. D'Aeth, staff officer of the group, Wing Commander C.J. Barclay, Group Captain G.C.E. Shilton and Wing Commander J.G. Davis. Together they pored over the charts in the map room at Coastal Command HQ and helped Brian to make his plans down to the finest detail for the south-western flank battle, which had to ensure absolute security throughout the invasion.

At this stage and right up to the day itself, the date of the invasion was and would remain top secret. As Brian later wrote in a newspaper article, he was frustrated at the lack of action or even interest from the top brass:

> From a meeting I attended as Air Officer Commanding no. 19 Group, Coastal Command, it was clear the Admiralty had no plan in mind. Having chewed it over for some time, they decided to ask for the opinion of the RAF representatives who, after all, were to do the job.
>
> We'd been working on a plan for some time, in conjunction with special groups of the Navy, based on our joint operations against the U-boats in the Atlantic and the Bay of Biscay over the previous two years. Roughly, this had involved the close co-operation between aircraft of Coastal Command and small hunting groups of frigates and destroyers. During that period we sank a lot of U-boats ourselves, using depth charges. ... Realising the forthcoming battle would be fought chiefly at night, we put in intensive training with the naval hunting at night during the two months before D-Day.
>
> The plan was put to the Admiralty as follows:

First of all to spot any U-boat putting its nose out of a port and heading for the Channel. To do this, Coastal Command aircraft would fly in oblong blocks – two planes to every block, beginning off the Bay coast on the night of 5th June – the night before D-Day. The patrols would then work up the Channel, according to the time it was estimated it would take the U-boats to travel. Backing up these patrols would be naval hunting units at strategic points in the Channel.

We aim to patrol every area of water every twenty minutes.

After questioning us carefully on our plan, the Admiralty seemed satisfied that it was satisfactory and likely to work. So it was approved.

Squadron Leader M.C.D. Wilson and Flight Lieutenant A.S.L. Robinson's book, *RAF Coastal Command Leads the Invasion*, provides an in-depth study of all aspects of Coastal Command and its expert personnel. He describes Brian as:

In command of the anti-U-boat group of Coastal Command, which was most actively engaged with the flank defence of our invasion forces from these underwater craft, was Air Marshal B.E. Baker, CB, DSO, MC, AFC. He is dark and stocky, and a former World War 1 pilot, who was educated at Haileybury, where he was a school-fellow of Air Chief Marshal Sir Trafford Leigh Mallory and Air Marshal Sir John Slessor.

Baker is a crack shot and an expert angler, with a wide knowledge of aircraft and a specialised experience of the peculiar nature of the Command's work.

Known as 'Bee-eee' to everyone in his Group, he never flaps, but maintains a calm of concentration with a quick, original mind and an enterprising outlook. His initiative has been proved in the field of battle. It is mainly due to the efforts of the Group.

As the chosen day approached, Brian and his team went over and over their plans, starting with making sure that every man and every machine would be in the correct place, with all the back-up and contingency plans, which had to be watertight. Still the date of Operation Cork was a secret to all but essential personnel, though the enemy hierarchy were suspicious that something might happen. However, they could not have foreseen the immediacy and scale of what was about to hit them.

To keep things 'normal', Brian returned to his RAF command at Plymouth whenever he could. The only people at Plymouth who knew the date of Operation Cork were the commander-in-chief of the garrison, Brian as commander of No. 19 Group Coastal Command, and his deputy.

One day, the Lord Mayor of Plymouth, Lord Astor, walked into the garrison commander's office and announced that the city's 'Salute the Soldier Week' would be inaugurated on Saturday, 3 June by Admiral Lord Chatfield. He had in mind a procession through the streets of Plymouth. Brian and the garrison chief exchanged glances, stunned by this announcement. What to do next? Before the chief said anything that might give away the secret, Brian, feigning enthusiasm, calmly agreed.

'Yes, we'll put on a good parade for you.'

That Saturday was to be the day when Brian and all the combined Allied air forces and support groups would travel to Portsmouth and other locations to take up their positions for Operation Cork to begin. Brian began to think up a credible plan for the day that would not involve any of his air personnel … Meanwhile, it was time for final preparations, and sharing the top secret plans was kept to the last evening.

Squadron Leader Hector Bolitho later wrote an article in which he described the evening in early June when he walked into a cinema on a cliff in Cornwall with other members of four Coastal Command squadrons. They numbered over 1,000 airmen, and along with the ground crews, they were tightly packed and very warm. However, sensing something important was about to happen, nobody wanted to leave.

After a short lull, the air vice-marshal (Brian) walked up to the microphone and started to speak, telling them all about the task ahead:

As everybody in the South of England must now be aware, a most daring and highly organised movement of troops and equipment, which has taken months to prepare and to perfect, is very shortly to take place, in the shape of an invasion of the Continent. This enterprise must not fail, whatever the cost. If it does fail, the war will be considerably lengthened, but if we succeed it will be considerably shortened.

Many lives and vast quantities of equipment will depend upon the efforts, not only during the passage of the ships in the initial assault, but during the period of build-up of the bridgehead afterwards.

Every ship in every convoy is carrying some vital piece of the whole – a piece without which the jigsaw may break apart.

A few ships sunk carrying these vital pieces may make the difference. The enemy will, in fact must attack these convoys with everything he possesses – destroyers, frigates, U-boats and super-fast E-boats. It is these against which you will be called upon to operate, if and when the time comes.

There are four Narvik frigates and one Dutch destroyer to be accounted for, as you have already dealt a very shrewd blow against the one and only Elbing left in Brest, which put her into dock … and dry dock at that.

The destination of the Narviks is of vital importance and we must do our best to get them out of the way early in the proceedings.

There does not appear to be much danger from E-boats, with the present dispositions, but I hope there will be a chance to show the Navy what the Air can do on a moonlit night, with bombs and rockets, if we can have an area free of ships to work in, so that you can get some value out of the training which you have been going at so hard.

The area at present looks as if it will be south-east of Start Point.

The remaining squadrons in the group will be attending to the actions of the U-boats. There are approximately 50 U-boats in the Bay of Biscay ports, as far as is known, so about 35 might be expected in the Channel. To cope with this, we propose to jam a 'cork' of Liberators, Wellingtons, Halifaxes and Sunderlands into the entrance to both the English and Bristol Channels, in an endeavour to keep these U-boats out, and we shall hunt them there to exhaustion.

Outside the convoy routes will be patrolling Avengers by day and Swordfish of the Fleet Air Arm by night.

To help us carry out this task, A.D.B.G. have provided adequate fighter cover but at the moment there are no enemy fighters to worry about. No. 10 group have searched for them in vain for some weeks, and what there are will I think be more engaged elsewhere.

In addition to which, aerodromes and radar stations will be attended by bombers and fighters.

That is the set-up and it is up to the aircrews to keep the air cover going as incessantly as possible, and for the ground staff to provide the aircraft for them to fly in the quickest possible time. By that, I mean a really quick turn round after landing, so that the aircraft is ready to go again if necessary.

Aircrews must encourage their ground crews by telling them what is going on and what they did in the way of fighting or sinking boats, so that they can all enter into this great enterprise as a squadron team. Aircrews must also make rest their first thought on returning from sorties, so that they also are ready to go again.

Aircrews got very tired in the Battle of Britain. You may also feel tired before this battle is over, but you must keep it up.

I feel confident that the U-boat is in for some very rough handling from ourselves and from the Navy, and that we shall get the better of it, as we did in North Africa.

Brian paused and looked round the room. He knew so many of these men and they knew him, so what better could he end with to ease the tension but a cricket analogy?

It may interest you to know that their number one batsman, who was attempting to play an innings in the Channel, was, to the best of my knowledge, sunk within a few miles of turning the corner round Ushant, a few days ago. Since then we have not seen or heard of the next man in. We got their first destroyer and their first U-boat. So far, so good. We can and must keep them out of the Channel so that every ship lands its cargo on the other side.

Brian then explained to the officers the logistics and destinations for the next day's movements to Portsmouth and other airbases. 'These four squadrons are but a small part of the force of naval and aircraft that will patrol the seas,' he reassured them. Finally, Brian emphasised the crucial need for secrecy for these last twenty-four hours before Operation Cork would go live.

First thing the next morning, 5 June, the mayor's special day began with a reception, which Brian was expected to attend with his wife. They duly arrived and mingled with the councillors and guests. Ten minutes later, as Brian checked the time, he spotted Jaimsie in deep conversation with a woman he didn't know. He asked an official who she was.

'Ah, that's Miss La Grande,' he said. 'I don't think she was invited. We believe she's a spy.' So saying, the official made a beeline for the intruder and led her away, while Brian joined Jaimsie.

'Did you tell her anything?' he asked.

'No, she had only just joined me and we talked about the weather, then our children.'

Highly relieved, Brian steered Jaimsie out of the building and into the bustling crowd outside, watching the naval parade marching past and waving to the people lining the streets. Little did the citizens of Plymouth realise that the smiling men and women in smart uniforms marching down their streets and returning their salutes were the handymen, butchers, cooks, cleaners and gardeners from the garrison, all in borrowed uniforms.

Wanting to get away, Brian took Jaimsie to the other side of the city. 'I'm going to show you a sight you will never forget,' he told her. He was right. The whole bay was full of a huge naval flotilla, protected by the landscape and sheltered from enemy reconnaissance. Within the hour, Brian was back with his squadrons, doing the rounds and keeping up morale. As everyone readied themselves and their equipment, Wing Commander Charles Bray found a few moments to write an article for the local newspaper, to be delivered after D-Day. In it, he set out his impressions of the situation and his trust in their commanders:

> The Germans will make every effort to cut and disorganise the supply lines of the invasion forces when the great combined operation takes place.
>
> Already there are unmistakable signs. There has been a lull in U-boat activity, undoubtedly due to recalling most of their craft to their bases around the Bay of Biscay, in the belief that the U-boats are in all probability waiting to strike from these bases.
>
> Coastal Command of the RAF has its plans ready.
>
> The Air Chief Marshal has at his disposal a complete Air Force of Liberators, Fortresses, Sunderlands, Halifaxes to attack the U-boats and Beaufighters to attack enemy shipping.
>
> Two men will shoulder most of the responsibility. Maker of the overall plans is Coastal Command's Chief Air Marshal Sir Sholto Douglas KCB, MC, AFC, and the other man who will play a very prominent part in carrying out the plans is the commander of the very important group, Air Vice-Marshal B.E. Baker, CB, DSO, MC, AFC.

That afternoon, all the pilots and aircrews flew across land to Portsmouth to join the massing troops and other services, while the ground crews travelled in trucks with their tools and supplies. The logistics of a gathering and finding spaces for the thousands of servicemen in their squadrons, plus feeding and watering them, had all been part of Coastal Command's plan and it worked smoothly.

As dusk fell on the eve of D-day, Brian and all the other Allied forces' commanders, wherever they were, gathered their troops and read out a message from the American President, Dwight D. Eisenhower:

Soldiers, Sailors and Airmen of the Allied Expeditionary Force,

You are about to embark upon the Great Crusade, towards which we have striven these many months. The eyes of the World are upon you. The hopes and prayers of liberty-loving people everywhere are marching with you. In company with our brave Alllies and brothers-in-arms on other Fronts, you will bring about the destruction of the German war machine, the elimination Nazi tyranny over the oppressed peoples of Europe, and security for ourselves in a free World.

Your task will not be an easy one. Your enemy is well trained and equipped and battle-hardened. He will fight savagely.

But this is the year 1944! Much has happened since the Nazi triumphs of 1940–41. The United Nations have inflicted upon the Germans great defeats in open battle, man-to-man. Our air offensive has seriously reduced their strength in the air and their capacity to wage war on the ground. Our Home Fronts have given us an overwhelming superiority in weapons and munitions of war and placed at our disposal great reserves of trained fighting men. The tide has turned! The free men of the World are marching together to Victory!

I have full confidence in your courage and devotion to duty and skill in battle. We will expect nothing less than full Victory!

Good Luck! Let us beseech the blessing of Almighty God upon this great and noble undertaking.

Dwight Eisenhower, President of the
United States of America

Meanwhile, Brian, his squadrons, Coastal Command and the Allied forces bedded down wherever they could for half a night's sleep, and the dark sea seemed to hold its breath …

Chapter Thirteen

1944

D-Day

At exactly 11 p.m. on 5 June, the eve of D-Day, the Allies simultaneously launched intensive jamming of German radar frequencies, which blinded the enemy's radar network from Cherbourg to Le Havre. This meant that Germany's early warning system was unusable in those precious early hours of Operation Cork. They could not detect Allied shipping or aircraft, or be immediately aware of Allied troops' movements. In addition, they could not forewarn U-boats of potential attacks or update their own land armies.

Perhaps this, together with how well the date had been kept secret, was at least part of the reason why, on the dawn of the crossings to Normandy, not one U-boat appeared from any of the French ports. In fact, it wasn't until the afternoon of D-Day that about forty U-boats poked their noses out of their sheltered pens in the Bay of Biscay and ventured out into the open sea. What they found must have been quite a shock – swarms of Allied aircraft patrolling the coast and ready to sting. Consequently, the U-boats, ten times slower underwater than above, had no option but to surface and attempt to escape their predators, no doubt in hope of doing some damage to the Allies with their cannons.

However, Coastal Command, in partnership with the Canadian and American air forces, were entirely focused on ensuring that as many U-boats as possible were disabled or destroyed, to allow the Allies' invasion troops – who were still disembarking from their landing crafts on the Normandy beaches – to land as safely and as swiftly as possible, unhindered by U-boats anywhere near the Channel, as well as to defend the Allied nations' coasts. These were the essential goals of Brian's Coastal Command group. Consequently, Allied aircraft were on constant patrol, looking for U-boats and dropping depth charges on them, if still submerged, or passing on their positions to Allied shipping to shoot them. As Brian wrote later in a newspaper article:

Time was vital and at their submerged speed they had no hope of intercepting the invasion fleet.
    Day and night they were harassed and pounded from the air.
    On the night after the invasion began, in conjunction with the Navy, we sank 11 U-boats – not a bad start.

On the morning of the second day, to protect some Coastal Command convoys bound for Normandy, some fighters of the Allied Expeditionary Force helped the navy to lay a gigantic smokescreen across the water. Meanwhile, there was scarcely a square mile of sea that was not under surveillance from one or other of the Command's anti-U-boat aircraft.
    Fighting was flat-out on the second afternoon and especially at night, and again it was mostly in the Allies' favour. Many U-boats were sunk and many others seriously damaged. Those that escaped destruction limped home. At night, most of the U-boats were illuminated by a Leigh Light and flares. According to Wilson and Robinson:

Many of the targets may well have been more seriously damaged than could be claimed on the insufficient evidence sighted. No U-boat fleet could stand such losses and no men could penetrate such a barrage of depth charges. ...
    The U-boats changed their tactics, but still they came out to fight. This time, instead of travelling above surface, they tried to pierce our air barrier below the water Our aircraft found only periscopes and oil slicks to aim at and, of all the targets in warfare, these are probably the most fleeting and the most difficult to attack.

One Coastal Command pilot, with the help of his crew, made a record on that second night that was not surpassed or even equalled by the end of the war. The pilot was 21-year-old Flying Officer K.O. Moore from Vancouver, Canada. He was the skipper of a Liberator in one of Brian's Coastal Command squadrons. Just before he came out on this sortie, Moore's squadron leader had jokingly told him: 'There are bags of U-boats around. You ought to sink two in an hour.'

Grinning, Moore replied, 'We'll sink two in half an hour tonight!' This is how the story was told:

Everything was quiet when 'G for George' approached the French coast at the beginning of the patrol. Things started to liven up … and clearly silhouetted ahead of us, fully surfaced was a U-boat.

As we approached, we could see that the U-boat crew had been taken by surprise.

About eight German soldiers on deck, apparently in utter confusion, were running like hell to man the guns. The U-boat commander however made no attempt to order his vessel to crash-dive and those manning the guns waited for us to come into close range.

Al Gibb, our front gunner, opened fire when we were within 1,000 yards and scored repeated hits on the conning tower and deck.

Simultaneously the U-boat's crew opened fire on us. I took evasive action, while Gibb continued to blaze away and then tracked over the conning tower.

The flak had been silenced during the last yards and as we released our depth charges I saw one of the crew in the 'bandstand' double up and fall overboard into the water. Warrant Officer Griese, in the mid-upper turret, and Flight Sergt. Webb in the rear turret reported a perfect straddle.

The U-boat seemed to jump into the air and explode, splitting wide open. I made a steep turn. On the water I could see wreckage and large patches of oil. As we stooged around for another five minutes we saw black objects in the water. Probably bodies.

Proof of the sinking of the U-boat was obtained by Flying Officer Ketcheson and Flight Sergeant Hamer, who took photographs that, when developed, would show the depth charges falling on either side of the submarine's conning tower.

Five minutes later, Flying Officer Moore took his Liberator back out on patrol over the Channel:

As we resumed patrol, Mike Werbiski was busy at the wireless set, sending out a flash report to Coastal Command that we had just made an attack. Don Griese came and said: 'Come on, let's get two U-boats.'

I remember telling him to be patient because the night was still young.

Almost immediately, Warrant Officer McDowell, the Scottish navigator, who at the time was adjusting his bomb sight, shouted out a warning that he could see another U-boat ahead, travelling fairly slowly. It was a small U-boat and it remained fully surfaced, making no attempt to avoid a fight.

Once again, Flying Officer Gibb opened fire from the nose-turret as the Liberator closed in for the attack.

This time, Flying Officer Moore and his crew found themselves up against much heavier enemy ack-ack.

The U-boat sent its flak in the shape of a fan. Moore flew right through it for the depth charges to be released as the Liberator roared over the conning tower at low level. A perfect straddle was obtained once more and the aircraft escaped being scathed. The U-boat was still on the surface after the attack, although it was listing heavily to starboard. Not one enemy shot had hit the aircraft. Moore continued:

As we circled, I saw the bow of the U-boat rise twenty-five feet out of the water, at a seventy-five-degree angle, and slide back into the sea. We made a third circuit over the spot where the U-boat had vanished. Here dinghies were in the water with three or four German sailors in each.

Then we went on with our patrol.

Flying Officer Moore paid tribute to his crew: 'The only one who showed any excitement was myself, after the second attack, when I exclaimed: "My God – it's sunk!"'

Kayo Moore, the name by which he was known in his squadron, was an unassuming young man. Now, this unassuming young man was awarded an immediate DSO for 'Exceptional courage and outstanding devotion to duty'.

Three other members of his crew were also decorated immediately. DFCs were awarded to Warrant Officer W.R. Foster of Guelph, Ontario, and Warrant Officer T.J. McDowell of Kilmarnock, Scotland, while Flight Sergeant Hamer of Colchester was awarded the DFM.

As Brian tells us:

There is a postscript to the telling of this successful mission. The crew of this Liberator were a mixed 'Empire' crowd, comprising seven Canadians, a Scotsman, a Welshman and an Englishman, but they all shared one superstition.

They wouldn't fly on any sortie without 'Dinty', their stuffed panda mascot from Montreal.

Dinty was there when they had fought with Heinkels, Nazi destroyers, when they attacked their first U-boat and again when they destroyed the two U-boats within twenty minutes. Dinty wore a Royal Canadian Air Force battledress, with the insignia of a warrant officer on his sleeve. Dinty's proudest day was when he was awarded the DFM by the Canadian Air-Force HQ.

That evening, when K.O. Moore got back to the officers' mess, he was cheered long and loud for his and his crew's unique record of dispatching two U-boats within the space of twenty minutes. Finally, the enormity of it all hit him. Half an hour later, wanting to congratulate the hero,

Brian searched the mess and finally discovered K.O. Moore 'slumped half in and half out of a comfy old armchair, so deeply asleep that he hadn't noticed the Mess dog happily chewing holes in his trousers'.

On 9 June 1944, there was a front-page headline proclaiming that the Allies had won the war in forty-eight hours, although this was a clear case of hyperbole. It was in fact wrong – but it was also right. It was wrong in that the war would rumble on for a few more months yet, but right in that at the end of forty-eight hours, Operation Cork – the defence of the landing beaches – had fully achieved its aims. Not a single U-boat managed to reach the Channel. Indeed, on the third morning after D-Day, HQ's radar detected that the few U-boats that ventured out were mostly heading east instead of west – the wrong way, which suggested their crews had had enough.

Somehow, on his return to Plymouth, Brian found time to write a letter to his sister Winifred:

Plymouth
11 June 1944

My dear Win,
Thanks for your letter. I am sorry that poor old Aunt Pen has had to be removed in the end, but it was obviously the only thing to do. Anyway, she would know nothing about it.

Well, we've finished our great effort at last, thank goodness. Having known the date and place for some weeks, I was mighty glad when it happened.

I was sure it would be too difficult for some people to keep the secret, especially when they heard somebody talking about what they were going to do on such a day, and you knew perfectly well they wouldn't. They might even be hundreds of miles away.

Sooner or later, quite by accident, someone might find it too difficult not to say anything, but it seems that nobody did.

It seems to be going very well – our part of it anyhow. We've given out some good wallops to a large number of the devils, and a good few won't come back any more.

Other than the operations room I have not seen much of anywhere else since we started.

Thank you for the telegram. We went out to Moorland Links last night, for dinner and dance and we had a good party. I managed to win a lovely lot of flowers and chocolates and I also managed to get three pairs of stockings for Jaimsie, which will please her very much.

The 'secret weapon' doesn't seem to have appeared yet. I thought the thing was going to whiz to London five minutes after we got going. In fact we were expecting one to arrive in here. For some reason they don't seem to recognise the importance of the place, and we were left out. It would have been a good target for them if they had arrived on Saturday, the day before the regatta. I've never seen so many ships and balloons in my life. They steamed along the coast all day and the whole of Plymouth was looking on. I should think they must have been a bit suspicious that something was up. It was most amazing how the waving flags and cheerings as the craft went out. I had a terrific salute when the staff paraded out on Saturday. It really was a terrific parade.

I must go and get on with some more administration.

Affec. Yrs.

Love to all,

Brian

Once all the Coastal Command administration had been completed, Brian was in demand again for voluntary work, starting with an inspection of the Plymouth Air Cadets at the Royal Citadel. On 24 July, Brian was surprised by a foreign envelope, forwarded to him by The Admiralty. When he opened it, he discovered that he had been awarded yet another honour. This time it was from the Republika Československá. Brian was now 'Grand Officer of the White Lion'. He was particularly pleased to receive this surprise honour as Czechoslovakia had been a loyal ally in Operation Cork, fighting bravely alongside Britain, Canada and the USA.

Six weeks later, Paris was liberated, followed gradually by the rest of France and parts of Europe. A few days after the end of the operation, Brian was introduced to the Chief of Staff, Viscount Portal and Field Marshal Lord Alanbrooke, who described Coastal Command's Operation Cork as a '100 per cent success' and congratulated Brian on his leadership of such a magnificent achievement.

On 15 August, Brian was awarded the honour of becoming a Knight Commander of the Most Excellent Order of the British Empire (KBE) for 'distinguished services in his operations in Normandy'. As the title suggests, Brian was awarded a knighthood. On the same day, he received congratulatory telegrams from both Lord Derby and Lord Alanbrooke. Sir Brian's investiture took place, as was the usual custom, in Buckingham Palace, giving him the title of Air Vice-Marshal Sir Brian Baker.

A few days later, Sir Brian welcomed a royal visitor to RAF Coastal Command – the Duke of Gloucester, Governor General Designate of Australia. The duke, in military uniform, took a genuine interest in the D-Day aircraft as Sir Brian showed him round.

On 3 September, Prime Minister Winston Churchill sent a telegram to Sir Brian to be circulated or read out to all who served in Coastal Command:

I send to you and to all your officers and men my congratulations on the splendid work of Coastal Command during the last three months. In spite of the hazards of weather, and in the face of bitter opposition from the armament of enemy U-boats and escort vessels, your squadrons have played a vital part in making possible the great operations now going forward in France, working in close concord with the Allied Navies, they have protected so effectively the host of landing craft and merchant vessels that the enemy U-boat campaign against them has proved a complete and costly failure. Many U-boats have been sunk or badly crippled in these operations in which squadrons of the Royal Air Force, of the Fleet Air Arm, of the United States Navy, of the Air Forces of the

Dominions and of our European Allies have all played their part. In addition, most effective attacks have been delivered against enemy shipping and very many hostile escort vessels and merchant ships have been sent to the bottom or heavily damaged. I know that the achievement of these fine results required that careful plans by Commanders and staff should be executed with the utmost skill and determination by the aircrews, who in their turn depend upon the tireless efforts of all who work for them on the ground. All have been united in carrying out a most successful summer's operations, of which you and your men may feel justly proud.

Undeterred by all his ongoing Command duties, with the war still raging in parts of Europe, Sir Brian continued to undertake his favourite voluntary activities whenever he could. However, his services were in demand, as the newspapers announced his new appointment:

Air Vice-Marshal Sir Brian E. Baker, KBE, CB, DSO, MC, AFC, will assume command of the Royal Air Force in East Africa with effect from November 15, 1944.

Chapter Fourteen

1944–1945

Victory!

Afdeal of new responsibility, but would also provide him with
intriguing challenges and opportunities, not to mention the very
welcome warmer weather. However, he would miss some of the great
chums he had worked with in the preparation and execution of such
a successful D-Day operation. One of these was William Hamilton,
who, on 24 October 1944, sent Sir Brian this telegram:

> Please pass to Air Vice-Marshal Sir Brian Baker, the Commander
> of the Fleet Air Wing seven, his staff and his many squadrons:
>
> I WISH YOU CONTINUED SUCCESS IN YOUR NEW
> ASSIGNMENT. It has been a pleasure and an honor to have
> served with your group. Good Hunting and Happy Landings.
> William H. Hamilton, Commodore, US Navy

Having said his goodbyes to Coastal Command for the time being,
Sir Brian flew himself across the Channel, which was calmer and less
crowded than the last time he had flown over it amidst the fighting
fury of D-Day, just six months before.

As he reached land again, Sir Brian flew in stages, across France
and the bright blue Mediterranean, to land at the RAF East Africa
Headquarters in Cairo, arriving on 15 November 1944 to take command.
Once he had met his new staff and taken a walk around the aerodrome,

he picked up some post and discovered that he had now been promoted to permanent Air Marshal – another accolade to add to his recent knighthood. Whilst he was pleased to receive this honour from his country, his immediate focus was on approaching his exciting new role with all his usual enthusiasm and vigour. Indeed, judging by an entry in his old school magazine, *The Haileyburian*, Sir Brian was fast developing a new enthusiasm, which beckoned him out on his first mini safari: 'Had a grand time in Nairobi. Drove out to the Game Reserve, saw hosts of animals of every variety at very close range. The best afternoon I've had for ages.'

And a few weeks later:

Still out in East Africa. Martially speaking, a dull spot, but full of every kind of sport and every kind of food. Hospitality grand, but exhausting (mind the waist-line!). Been shooting with C.H. Stockley (old Haileyburian), a great naturalist. He takes photographs for *The Field* and *Country Life*. I got chased by a rhino!

In between occasional forays out beyond the perimeter, Sir Brian spent most of his time getting to know his fellow officers and as many of his airmen as he could. But it didn't stop there. Ever since his first days at Montrose, three decades earlier, when he was learning to fly, Sir Brian had always taken a great interest in the aircraft themselves. He had often sought out the ground crews and mechanics and asked them questions. This hadn't changed, so he soon made friends with everyone at the Cairo HQ.

As Christmas 1944 approached, Sir Brian's thoughts turned to his wife and daughters, preparing to celebrate Christmas without him, yet again. With them in mind, Sir Brian knew that all the men and women working in or for the air force would have similar thoughts. It was the custom for the station commander to deliver a speech to all ranks at RAF East Africa Headquarters, but this year it was decided that it should be broadcast on the radio, so that all those working away from HQ could still listen to it on Christmas Day. Fortunately, Sir Brian was a confident writer and speaker, so the idea of being broadcast didn't worry him unduly. He made time to sit down and write his speech. He knew that the most important thing was to allude to loved ones at home, and what everyone wanted was some hope that next year the war would be over and families could be reunited for Christmas. Of course, Sir Brian couldn't promise that, but he could give all ranks hope that it might be possible to have a phased home leave programme reinstated relatively soon. So he began to write his speech with all these points in mind.

On Christmas Day, at 1920 hours on the dot, Sir Brian had the signal from the sound recordist and began to deliver his speech:

This is the sixth Christmas that we have spent under war conditions and very naturally we are beginning to feel the strain. I would however remind you that there is another nation who is spending a sixth war-time Christmas and what a Christmas it is going to be for them. Hemmed in on every side by vast and well-equipped armies, bombed and blasted by the greatest power ever known; supplies running low, run down by the Gestapo, all hope of victory gone and knowing that they are shortly to be subjected to unconditional surrender. What a helpless dawn for them in the year 1945.

The Japanese must also face 1945 with some misgivings. They are being pressed back slowly but surely by Burma. Their long lines of communication are very seriously menaced by two very powerful fleets and their own capital is already within range, yes and has already been bombed by American long-range bombers. That is the situation at the end of 1944.

The present, satisfactory state has been brought about in the first place by the dogged tenacity of the British people who, in the dark days, stuck out their chins and took blow after blow without flinching.

Secondly, the patient and loyal service of all ranks of our fighting forces, whether in the heat of battle or in more distant lands, remote from visible action. It is to those airmen and airwomen from the seat of war, far from their own families and homes, who, nonetheless play an integral part in the war, that I wish to speak.

Some of you have not seen your families and homes for several Christmas days. And nobody regrets that more than I do – the fact that tour-expired airmen have not been able to get home this year. Everything possible is being done to obtain reliefs and transport to bring about this happy event as soon as possible and I am confident that early in the New Year you will see your well-earned repatriation accomplished. You have indeed played your part well in the past year.

And now a word for the new-comers, especially the women in the auxiliary Air Force who are perhaps spending their first Christmas overseas. Africa is a vast continent, full of the most interesting things. Make the most of your spare time to explore this great country, the ways of living of the inhabitants and see the wildlife, with which it abounds.

Those who have gone before you have enjoyed great hospitality from this country and I know you will receive the same; and here I would like to take this opportunity of thanking all those whose kindness and generosity have done so much towards the entertainment and enjoyment of RAF men and women.

The spirit shown by the kind friends you have in East Africa and the islands is the spirit of Christmas – 'Good will towards men'. In your own associations with your fellow airmen and airwomen, let us in the coming year follow their good example and so try to make our own lives a little happier and brighter, day by day, through our own efforts and our own presence.

To all RAF men and women, members of the finest Air Force in the world.

I send my best wishes for a merry Christmas and a Happy New Year. We are going to see this job through, wherever we may be. We were called to serve, and see it through we will.

Sir Brian's speech was well received across the airwaves, as well as by all ranks at the Cairo base, especially the hope of home leave for all in the coming year.

Meanwhile, another airman, full of Christmas cheer (probably of the liquid variety), penned his own Christmas contribution, which was passed round his pals with much merriment:

'Erb's Christmas

There's another Christmas comin'
'an I opes you all 'as fun,
Although it's sorta funny
'Avin' Christmas in the sun.

I'll tell abaht a Christmas
It's the queerest I've 'ad yet
If I live ter be a 'undred
It's one thing I won't ferget.

We 'ad the usual dinner
All the lads were 'ere in force,
An' I'll give me own impressions
Though I didn't stay the course.

The CO made some speeches
An' we all began to cheer,
Though the adjutant got rattled
When an SP pinned 'is beer.

The Station Warrant Officer
Was crawlin' on the deck
Assisted by a sergeant
Wiv 'is arms arahnd 'is neck.

A shower of erks an' corporals
'ad pinched the CO's car
An' they drove it through the Naafy
Only stoppin' at the bar.

It didn't do the 'Ol Man's car
No Ruddy good at all,
A' still they point aht the places
Where it 'it the Naafy wall.

I looked up at the flag-pole
An' I noticed sumfink queer,
I 'ad to take another look
An' have another beer.

The strangest thing I've ever seen
Since I've been overseas,
Is a pair of Waafs' blue underwear
A floatin' in the breeze.

They said the MO done it
But they didn't 'ave no proof,
Although they found another pair
Up on the clock 'ouse roof.

Then just as we got goin'
Wiv a bonfire on the square,
I drank my fourteenth bottle
An' I passed out then and there.

I woke up in the mornin'
Wiv a funny sorta 'ead,
An' a man poured aht me breckfust
In a glass beside me bed.

Tork abaht the festive season
You can 'ave no ruddy doubt,
If the next one's 'arf as festive
Then young 'Erbert's going out!!!
T.M.L.S.

Sir Brian must have enjoyed this clever verse, especially the true references to his car, as he sent it home to Winifred to put in one of her albums, in which she had been documenting her brother's career and other activities since the First World War began.

In the first two months of 1945, Sir Brian went travelling to as many of the RAF outposts as he could across East Africa. For part of his travels, he took some of his fellow officers with him – those who were keen cricketers, like he was. He planned a tour that would fit in as much cricket as possible. He sent an account of his travels to his old school and it was included in the next issue of *The Haileyburian* newsletter, with an introduction by the editor:

A geography lesson – you'll want a map to follow Air Vice-Marshal Sir B.E. Baker in his recent, hectic ten days' trip of 5,600 miles around Arabia; an official visit to those far-flung RAF outposts of the Empire, on which (I understand) the sun never sets, though many out there must wish it did: Heliopolis, Habbaniya (Baghdad), Babel, Shaiba and Bassa, the head of the Persian Gulf, Bahrain and Shanja in the Gulf. Razall Radd, Masirah in the south-east corner, Salalah, Ryian, Aden, Port, Sudan, Wadi and Haifa.

The heat and humidity was terrific; our men at these lonely stops must have a rotten time – soft sand everywhere, no vegetation or livestock – they mostly live on fish.

At Razal Radd, for example, there is nothing, so everything, water included, comes by ship or air.

The editor added: 'One can be sure Baker's the man. He got back to Cairo just three-quarters of an hour after scheduled time.'

On his return to HQ, Sir Brian discovered that he had been elected to join the Cricket Committee. To celebrate this accolade, he had lunch with two distinguished old school friends: J. de Freitas ('the genial judge') and H.G. Vincent, who was Under Secretary of the Ministry of Civil Aviation and private secretary to both Stanley Baldwin and Ramsay McDonald.

Sir Brian's first cricket game after his travels was a 'major match' against 'Royal Gymkhana', in which he showed that had lost none of the talent that had made him the captain of the RAF cricket team all those years ago. The sports reporter wrote: 'Sir Brian Baker played a very fine innings, scoring most of the runs (67) on the off-side, with an occasional well-judged shot to leg. He laid the foundation for victory.'

Now, at the age of 49, Sir Brian was quietly pleased that he had maintained his prowess amongst lads less than half his age. It was a similar story when an Australian touring XI challenged 'Sir Brian's team' to a match – and lost by forty-eight runs.

As Commander-in-Chief of RAF East Africa, Sir Brian received an invitation from King Farouk to join him on 6 May 1945 at his Abdeen Palace in Cairo for a levee – a reception and party for important visitors. Sir Brian's invitation required him to be ready to be the first guest presented to the King.

When he arrived back at HQ, Sir Brian was handed an important telegram. There had been rumours over the past few days, but it must have been a wonderful feeling for him to open the small flimsy envelope and read that the Soviet Union had forced the Germans to surrender unconditionally and sign a peace treaty with the whole of Europe. The treaty was due to be signed the next day, Tuesday evening, so Wednesday, 8 May 1945 would be designated Victory in Europe Day, or VE Day.

This wonderful news spread across the base in minutes, and within half an hour, a party was being arranged. There were high spirits everywhere, along with feverish plans and preparations for food, drinks, decorations, games, music and activities. Meanwhile, Sir Brian received another message. This time he was to relinquish his command of East Africa and instead was to be appointed to take a new command as

Senior Air Staff Officer of the RAF in the Mediterranean and the Middle East. Fortunately, he would continue to be based at the Cairo HQ and therefore be able to enjoy all the victory celebrations with his men and the WAAFs.

Having completed all their preparations, the party organisers realised that they would have to spread the event over two days in the sergeants' mess. Programmes were printed and delivered – three pages of celebratory activities, all timed to the minute.

The first day began at 0900 hours with a 'Breakfast Special', followed at 1000 hours with a thanksgiving service, which must have been quite short as at 1030 hours, the bar was opened. At 1100 hours, there followed a novelty darts competition, a singsong and then lunch, with all food 'served in a running buffet'. The afternoon began with a radio news bulletin, and then there was a photographer to take group photos and one large gathering photo. After that, gramophone records were played until 1500 hours, when they switched on the loudspeaker for the radio broadcast of Britain's National Victory Celebrations. Then came a high tea, a running buffet and indoor sports competitions, including semi-finals and finals of darts. In the bar, table tennis in 'A' room, shove-ha'penny in the bar, crib in the ladies' room and chess in the upstairs alcove. 'Housey-housey, eyes down' with cash prizes attracted many competitors. This was followed by: 'Number 2 Mess bravely presents a play entitled *Crash Demobilisation, Round the Bend*, a play in three acts'. Finally, at 2030 hours there was a Grand Dance to end the evening.

The second day was in similar format, with various sports, including, of course, cricket on Gezira Island and all the prize-giving. There was also 'impromptu dancing', tennis, another 'spoof' play, the band, and a variety of other activities, as well as an almost all-day buffet and bar, with the concert party at intervals. After all that, everyone must have needed a third day to rest and recover.

A few days later, on 15 May, it was the commemoration of the Battle of Britain. On this occasion, the entire presence of RAF East Africa and the Middle East gathered for the formal remembrance of that 1940 battle. A Cairo newspaper published the following article, along with a photograph of the RAF procession, and with the text of the reading chosen for that ceremony:

> To commemorate the fifth anniversary of the epic battle which broke the power of the Luftwaffe, a ceremonial Colour-Hoisting Parade was held at El Alamein Club, Cairo on Saturday. 400 men and women of the RAF and the WAAF took part in the march-past, where Air Vice-Marshal Sir Brian Baker, AOC-in-Chief, Middle-East took the salute.

The reading was taken from the Book of Ecclesiasticus, Chapter 14:

> Let us now praise famous men: and our fathers that begat us. The Lord hath wrought great glory by them, through his great power from the beginning ... All these were honoured in their generations and were the glory of their times ... their seed shall remain forever and their glory shall not be blotted out. Their bodies are buried in peace, but their names liveth for evermore.

Sir Brian knew so many good men who had perished during the Battle of Britain, as well as D-Day, so this will have been a poignant ceremony for him. On 10 June, he received Poland's war decoration, the Commander's Cross with Star, investing him as a member of the Order of the Polonia Restituta.

Sir Brian spent most of that summer flying between the bases he commanded ... along with a handpicked cricket XI. As usual, the boys at Haileybury were encouraged to follow their journey: 'Sir Brian Baker has just made more history by taking on an RAF XI by air on a cricketing tour of East Africa.' Of course, Sir Brian had played some cricket in this region before, but it was rather ad hoc, between work visits, whereas this time it was all and only about Sir Brian's favourite sport. The editor of *The Haileyburian* continued:

They started in Aden, where the wind blew so hard that the bails wouldn't stay on.

Next was Nairobi against the Kenya Kanbories and secondly versus a USA XI.

Their third stop was Thilia, where they played Thilia's maintenance team on a matting pitch in the equivalent of a hayfield.

Next was Entebbe, where they played two matches: one against the local side and the other against the Uganda Kobs; both matches were played on the Equator.

The fifth stop was Khartoum against an XI captained by an old school friend.

On account of the heat, the hours of playing were 'rather trying' – up at 5 a.m. to play at 6.30–9 a.m.

When not playing cricket they were inspecting game reserves. They encountered sixteen species of game, including lion, elephant, water-buck, crocodile and a rhino, which allowed itself to be chased by Sir Brian and his team for 20 minutes (perhaps making up for the time Sir Brian himself was chased by one of its relations).

Other spare moments were occupied by dances, dinners and limitless hospitality. …

I am sure that under B.B.'s leadership, a good time was had by all. Wherever they went an RAF flag was left as a souvenir. It should be added that Sir Brian's XI won all their games, though this is of minor importance.

While Sir Brian and his team had been batting and bowling their way round East Africa, most of the world was adapting to their hard-earned freedoms and adjusting to new opportunities. However, there had remained just one country, Japan, still at war with the world, but the Americans were working on that and, on 2 September 1945, Japan finally had no choice but to surrender. For the first time in six long years, the world was at peace and looking forward to Christmas.

Chapter Fifteen

1946–1949

The Berlin Airlift

Sir Brian started off the New Year of 1946 with another unexpected honour. This time it was a second and higher decoration from Czechoslovakia – their Military Cross.

After so many years packed full of action and danger, Sir Brian had a much calmer, quieter year, continuing as Commander-in-Chief of RAF Mediterranean and the Middle East. As well as visiting RAF airbases across his command, one of his favourite roles was to visit and inspect aircraft carriers, landing on their decks as he used to do between the wars, when he trained sailors and airmen in how to operate them. However, now Sir Brian was visiting to inspect them and sometimes help address any problems on board. No doubt, notice of an imminent visit from Sir Brian caused apprehension and a lot of scurrying about the ship in preparation for his inspection, as one ship's log demonstrates: 'The colour-hoisting parade was held in the early morning and the Commanding Officer (of the ship) advised all ranks that the Air Commander-in-Chief would be visiting the squadron and station on the morrow.' Then, after the visit:

The main event of the day has been the visit of the Air Commanding Officer, the Air Vice-Marshal, Sir Brian Baker, KBE, CB, DSO, MC, AFC, who arrived by Catalina aircraft from Pamand, was accompanied by Group Captain, Lord G.N. Douglas-Hamilton, OBE, AFC. The visitors expressed satisfaction with all they saw and left in a Hudson aircraft after lunch.

As he often did when in England, Sir Brian not only kept in touch with his old school, but he also visited Haileybury whenever he could to attend sports events or present prizes, particularly for special occasions.

In the autumn of 1946, Sir Brian was invited to give the Haileybury boys an inspirational talk about his career. As he had already booked a month's leave over Christmas, he agreed. The day came in early December and Sir Brian had a full house, all with eager smiles, so he launched straight into his speech:

When war broke out in 1914, I found myself with a commission in the Rifle Brigade. I was, however, very attracted to aviation and in my battalion I found a kindred spirit in another officer who was mad keen to transfer to the Royal Flying Corps. I'm afraid we were both more interested in watching the flying at Joyce Green across the river than learning musketry at Purfleet.

By August of the following year, both of us had obtained a transfer to the RFC.

Like so many things though, it all went back to school days, and Haileybury had no small share in it. I was at school during a fascinating period in aviation history. Blériot had made his historic flight across the Channel the year before I entered Lawrence House in 1910. There I found as Housemaster, F.W. Headley, who was quite an authority on the flight of birds. He used to show us photographs of his pigeons in flight, taken from his study window.

Every week I read *The Aeroplane* from cover to cover. Its photographs of extremely daring flying feats encouraged my keenness to fly. And visits to Hendon Aerodrome finally determined me to try my hand at it when I left school.

One last factor was the Army Manoeuvres during the summer holidays of 1912. These were the first manoeuvres in which aircraft were used, and every morning I used to cycle out to watch the flying. Exciting days indeed they were. It was then that I first met the late Colonel Cody and Air Chief Marshal Sir Robert Brooke-Popham.

It was at Montrose I first learned to fly, in the very able hands of R.S. Maxwell (a former Hailey student). How different instruction was then than compared to nowadays.

One and a half hours dual in a Maurice Farman was all the training I had, and in less than twenty hours (only twelve of which were solo) I was over the German lines in a B.E.2c.

Aircraft have since developed, and training methods with them. The accomplished products of our training schools have proved second to none.

At the end of the war, I was a Squadron Commander. I realised there was no hope of retaining that rank. Neither did I mind very much. The grand thing was to be able to stay in with a permanent commission.

The inter-war years enlarged my experience far beyond my wildest schoolboy hopes. I was overseas in Ireland, Egypt and Aden. I commanded the Experimental Section of the Royal Aircraft Establishment at Farnborough. I was Chief Flying Instructor at Leuchars and Senior Air Force Officer in both HMS *Courageous* and HMS *Eagle* aircraft carriers. These were a few of my tasks. They were varied, they were all connected with flying and they all made for happiness and contentment.

The war took me from Iceland to East Africa and peace now finds me in the Middle East.

To some it may seem that the thrilling and exciting days of the early pioneers of flying, such as I grew up in, are over. I do not believe it. Development is always going on and the war has speeded it up.

For the next twenty years we shall be hard at work enlarging on the discoveries made during the war. There is a vast field of development – in aviation proper, and the numerous aids to flying which science is forever producing.

I have enjoyed my share of the pioneering days and the developments of recent years. Yet I could wish that I were once again about to embark on a career in the Air Force, in order to be in as a young man, working on the problems ahead.

It remains now for another generation than mine to conquer the sonic barrier, to carry jet propulsion further, and to keep for our country the lead in world aviation. And who shall do this but the Royal Air Force?

The year 1947 started quietly, with Sir Brian back in the Middle East. However, the government, under Winston Churchill, were now looking for ways to save money so there were rumours circulating about how that might be achieved. Meanwhile, in Europe, there were hints of possible troubles to come.

At the end of the Second World War, the four main partners in the victory had agreed to partition Germany between them. This was organised so that Britain, the USA, France and Russia took on responsibility for their own occupied territory, each with their own garrison. Berlin, too, was divided into four sectors. However, the capital city could not be so easily managed, as most of it was deep within the Soviet zone. The bone of contention was that Russia wanted all of Berlin for itself and perhaps later, the rest of Germany. It was all talk and belligerence for now, but the British government could see this as a potential problem in the future.

The RAF Transport Command had been one of the sections that the government was considering to reduce, or even disband, in order to save money. Now, they not only changed their minds, but agreed to reinforce it.

On 1 July, Sir Brian was promoted from air vice-marshal to air marshal, which must have pleased him, but as a man not given to self-aggrandisement, it didn't change him. He enjoyed his life in the RAF, whatever it might bring.

Just three months later, the Ministry called Sir Brian back to Britain to take on a completely different role from his Middle-East command. This time he was to become Commander-in-Chief of the RAF Transport Command. This certainly was a new direction for him, but he realised that his proven successes in organisation and logistics in the Battle of Britain, D-Day and other roles must have played a part in his being selected for this position. No doubt, he was soon briefed about the precarious situation in Berlin, and was introduced to his team. He looked forward to meeting his American counterparts, too. But first, he had been invited to have a personal audience with King George VI at Buckingham Palace. By now, Sir Brian was almost a friend of the Royal Family and enjoyed that afternoon's visit, answering the King's questions and having a general chat over a cup of tea.

Sir Brian took up his new command almost immediately and reviewed the briefings and updates from both Britain and America. From the latter, it was clear that the US Transport Command didn't think much of their British counterparts, the Royal Air Force. Nevertheless, Sir Brian refused to be rattled. He was determined that he would soon bring the Americans round and show them that from now on, the British Command was going to be every bit as capable as they were. He hoped that they would be able to come to trust each other enough to collaborate in preparing for whatever might be needed.

Indeed, it wasn't long before things started to escalate between the Soviets and the other occupying powers. At the time of partition, all four nations signed an agreement, but the Russians were now dissatisfied with that treaty, so they withdrew their co-operation with the other three nations and began to agitate against them. Sir Brian's reaction was to discuss with his team and explore all the potential scenarios. He then started to sketch out some preliminary plans for each. However, it was too fluid a situation to plan possible solutions when they didn't yet know what the problem was going to be.

Sir Brian and his US counterpart started planning together for all possible contingencies. In March 1948, Russia pulled out of the Allied Control Council, thereby cutting off any possibility of co-operation. On 24 June, the Soviets imposed a blockade, barring their erstwhile partners the use of all roads, railways and canals leading from the west to Berlin. This meant that a city of more than 2 million people had suddenly lost all access to food, fuel, medication and other supplies needed to keep their part of the city alive.

This was one of the most likely scenarios that Sir Brian had been working on in detail for several months and had shared with the American Transport Command. The two nations' co-operation was key to the success of this whole operation. Working together, they ensured the availability and procurement of essential goods and their transport to British aerodromes that were suitable for aircraft from American, British and French air forces. At the same time, the army and air force personnel of all three nations needed to be trained as quickly as possible and the aircraft themselves stripped out and adapted for this urgent task.

All of this was a mammoth undertaking – and it needed to be implemented immediately to avoid the Berliners' starvation.

So it was that, on 26 June, the Berlin Airlift began. Several different types of aircraft were commandeered – both military and civilian cargo planes, as well as stripped-out passenger planes, gutted to provide the most possible space for their cargoes. After a few days, it occurred to Sir Brian that instead of returning some of the aircraft empty, a few adaptations would enable any refugees, citizens or military personnel trapped in West Berlin to be flown to freedom in Britain or America. This proved to be a great success.

From 26 June 1948 to 30 September 1949, the Berlin Airlift:

- fed and provided heating and cooking fuel and medical supplies for over 2 million people every day for fifteen months
- provided raw materials to keep manufacture going for Berlin's economy
- brought out thousands of malnourished adults and children
- brought out Siemens and other goods to help sustain their economy
- made more than 1,500 cargo flights into Berlin every day
- landed one plane every forty-five seconds
- delivered more than 4,500 tons of cargo every day
- delivered more than 2 million tons of supplies over the duration
- made a total of 278,228 flights by USA, Britain and France combined

- notched up nearly 6 million flying hours
- exceeded 92 million flight miles.

Additional flights and deliveries were made by Australian (flying boats), New Zealand and West Indies air forces.

All flights came into West Berlin's only airport, Tempelhof, where the passenger building had to be demolished to provide more space to build three new runways and a larger area for unloading deliveries and organising distribution.

At the end of the Berlin Airlift, all parties agreed that it had been a great success and, most pleasing for Sir Brian, was the American commander's judgement that although he'd had misgivings at the outset, British plans, logistics, leadership and co-operation, together with it its air force, army and support staff, had been instrumental to the success of the whole operation and, most of all, the British planes had been 'demonstrably indispensable'.

On 7 December 1949, Sir Brian and all the men and women of Transport Command, together with other RAF and army personnel – all those who had any part in the Berlin Airlift – were invited to Buckingham Palace to take part in a parade. The *Daily Telegraph* featured a photograph of the parade and described the occasion as follows:

NATION HONOURS AIR LIFT MEN:
PALACE PARADE
THANKS FOR CONTRIBUTION TO WORLD PEACE

Men and women who operated the Berlin air lift were honoured by the King, the nation, London citizens yesterday when 260 representatives of the organisation went to Buckingham Palace and were later the guests of the City Corporation.

They had an enthusiastic reception during a march to Guildhall, by way of the Mall and Trafalgar Square, and at Guildhall, Mr. Henderson, Secretary for Air, referred to their contribution to world peace.

The parade was commanded by Air Commodore J.W.F. Mercer, who was deputy to General E.W.H. Tunner, Head of the Anglo-American Air Lift Task Force, during the operations.

The flights into which the parade was divided included RAF air and ground crews, WRAF and headquarters control staff, Transport and Coastal Commands and representatives of the Commonwealth air forces. There were also contingents from the British Army of the Rhine, the United States Air Force and Naval Air Force, the British airline corporations and charter companies and representatives of welfare organisations.

They paraded in the inner quadrangle of Buckingham Palace. In the uniform of a marshal of the RAF, the King was accompanied by the Queen, who wore a coat and hat of Air Force blue. They reviewed the parade and a Royal Salute was given. As the King and Queen walked along the ranks, they paused frequently to talk to men and women, including Americans.

The article then described how two RAF bands led the parade and all the top brass between throngs of cheering onlookers lining the streets. Businessmen, typists, caretakers and other city workers poured out to swell the crowds and stop the traffic, all wanting to join the celebrations. Even policemen doffed their helmets in praise. When the parade reached the Guildhall, the Lord Mayor, sheriffs, aldermen and other dignitaries stood in their regalia to welcome them and led them inside for a civic reception and luncheon, where the Lord Mayor raised his glass, proposing the health of the RAF, describing their Berlin Airlift operation as 'one of the most brilliant in the annals of the RAF'.

A few days later, an anonymous woman delivered to Transport Command's HQ a large gold cup with a label bearing the words: Thank You for the Berlin Air-Lift.

With all the congratulations and ceremony for the triumph of the RAF Transport Command, Sir Brian was proud of his team and all their hard work – all the more so when he received the USA's award of The Commander of the Legion of Merit USA for his excellent co-operation with the American Transport Command. He told his colleagues what he genuinely felt, that this medal was an honour for them as well as for him.

Sir Brian could have supposed that his and their success would have secured the government's assurances that they would continue to support this beloved Transport Command ... but no such assurance was forthcoming. Understandably, Sir Brian was disappointed by the uncertainty of funding and support. Indeed, he felt that not only were they undervalued but there was also a sense of abandonment, even betrayal. It is not hard then to imagine how Sir Brian must have felt when he saw on the front page of the *Daily Telegraph* on the news stand out on the street for everyone to see:

AIR TRANSPORT COMMAND MAY BE ABOLISHED
EXTENT OF ECONOMY CUT UNDER REVIEW

Although Sir Brian knew this was a possibility, nobody had warned him that it would be in the papers so soon. He wasn't even given the chance to forewarn he is team. Ironically, elsewhere on the front page was a photograph of Princess Elizabeth and the Duke of Edinburgh being welcomed by a smiling Sir Brian Baker on their return from Paris. (As Commander-in-Chief of Transport Command, Sir Brian always attended when any of the Royal Family departed for or returned from foreign shores.)

The news article continued the story:

The abolition or drastic reduction of the RAF Transport Command is being considered as part of the cuts in defence expenditure, announced by Mr. Attlee on October 24th (just a month after the end of the Berlin Airlift).

Transport Command has shrunk to a shadow of what it was at its peak in October 1945. It is now commanded by Air Marshal Sir Brian E. Baker and consists of two groups.

Sir Brian scanned the rest of the article and was shocked to find that there was no mention at all of the Transport Command's great work during the Berlin Airlift.

Sir Brian's family still remember how disappointed, even disillusioned he felt. However, as a lifelong optimist, he put forward the case for keeping the Transport Command fully staffed and ready for any future needs. When that didn't work, he talked it all through with Jaimsie and perhaps with his sister Winifred, too.

He was now ready to make the most difficult and yet easiest decision of his life … It was time for a new adventure.

Chapter Sixteen

1949–1979

The Last Hurrah

S ir Brian sat down at his desk and wrote his letter of resignation from the Transport Command as from the end of March 1950. He also requested that his name be removed from the RAF Active List from that date. He had given thirty-five years of his life, first to the Royal Flying Corps and then to the RAF. Now he sat back and relished the thought of all the opportunities that were open to him.

His first project was one he'd been planning in his head for quite some time. No prizes for guessing that it had something to do with cricket! He wanted at the same time to thank and raise the spirits of his group at Transport Command who had worked so tirelessly with him during the Berlin Airlift, saving 2 million lives. Sir Brian and his staff were all due a lot of leave, so it was the perfect time. He explained his idea to them, which met with great excitement. Out came the maps and the group huddled round while Sir Brian pointed out the places they could visit for matches on their tour across three continents. Of course, there were some staff whose home commitments prevented them from taking part in this grand adventure, but they agreed to man the fort during their colleagues' absence.

Sir Brian's plan was to visit all the commands and outposts he had held in his charge since D-Day. Some of the destinations would be previous cricket grounds, from professional club pitches to parched hayfields. Once the matches had been arranged, routes mapped out and accommodation finalised, they were ready to go. The press discovered Sir Brian's ambitious agenda just before they left:

STRENUOUS AIR TOUR

Air-Marshal Sir Brian Baker, C-in-C Transport Command, is leaving in a fortnight on an inspection flight of the trunk routes to the Middle and Far East. Such a route presents a very strenuous month's work, which will be carried out to a timetable of RAF exactitude.

The Air Marshal drives himself hard and his staff work to the same energetic schedule.

Double crews are carried to man the aircraft, usually a York is employed on a trip of this kind.

An occasional day off, however, is allowed, for Sir Brian likes to take it where he can have a game of cricket.

It is said that in the personnel for the crews, airmen with cricketing prowess are not overlooked.

Like another Air Marshal, Sir Arthur Sanders, Vice-Chair of the Air Staff, Sir Brian is an old Haileyburian. Both naturally are too young to have been the Prime Minister's contemporaries at Haileybury.

The C-in-C Transport Command will miss seeing Sir John Slessor, the CAS designate (also an old Haileybury friend of Sir Brian's), now on tour of the RAF overseas. I gather that their planes will pass one another over the Indian Ocean and exchange salutes.

Sir Brian did indeed make sure that he didn't overlook any of his staff who had cricketing prowess. He could only take two aircraft on his officially sanctioned 'inspection tour of East and West Africa, the Middle East and the Far East' (starting and finishing in England), so he took with him eleven staff officers and two aircrews. Sir Brian piloted one plane and another D-Day veteran piloted the other.

Their first destination was Cairo, where they had a very warm welcome and a good match on the RAF cricket pitch. From there, they toured several outposts, large and small, across East and West Africa. They were cheered everywhere they went and in some places they received military or ceremonial greetings whilst at other destinations they were welcomed with luncheon and a beer, or a cup of sandy 'Yorkshire tea'. However, the most impressive welcome must have been in Saudi Arabia, where they were met by the Sultan of Muscat and Oman and his bearded troops, in their turbans and white robes, wearing heavy cartridge belts and carrying rifles and ornamented daggers. However, the smallest outpost of all excelled in the warmth of their welcome and their hospitality. This was the remote RAF station on the volcanic Masirah Island, 1,000 miles north of Aden and 40 miles from the coast of Oman, near the south-eastern tip of Saudi Arabia. Here, after sunset, a group of native islanders lit a bonfire and staged a medley of folk dances for their visitors.

Before leaving next day, Sir Brian learnt from the commander, Flight Lieutenant Jones from Kidderminster, that this small, most isolated RAF station island of Masirah had no water, vegetation or food. Water for all purposes had to be converted from seawater and everything else had to be brought in by sea or air. Ironically, Sir Brian discovered that all these supplies were brought in via the other group of Transport Command, also threatened with cuts or abolition. After touring the rest of the Middle East, Sir Brian and his staff flew east to stations in Asia for a few more matches before they headed back home.

On their return, Sir Brian's group was in the newspapers again. This time it was to congratulate them on such a successful tour. One of the sports journalists pointed out that whilst the recent eleven-month Test series between England and Australia had been the longest cricket tour in duration, Sir Brian's Transport Command 'World Cricket Tour' had broken other records – notably, the number of matches in the shortest time, and the number of destinations in one tour. Indeed, the sports press hailed it thus:

Perhaps the most spectacular cricket tour ever, was that arranged by Air Marshal Sir Brian Baker. In 1949 he made an official tour of his World-wide Transport Command. The party was away 28 days, covered 23,000 miles and some of the places they played in were: Ismailia, Colombo, Changi (Singapore), Aden, Khartoum, Nairobi, Entebbe (Uganda) and Takoradi (Gold Coast).

Back at his desk at Transport Command HQ in Teddington, Sir Brian caught up with all the paperwork and visits to various functions. Attending an exercise at Old Sarum, he noticed that seven old Haileyburians were present, so he found a photographer willing to take a group picture of them all together for their old school's magazine. The seven standing in a row were: Air Marshal Sir Brian Baker, Air Marshal Sir Arthur Saunders, Prime Minister Clement Attlee, Air Vice-Marshal G.H.A. Piddock, Sir Brian's old friend Air Chief Marshal Sir John Slessor, Group Captain H.M.A. Day (former prisoner of war), and Air Marshal Sir William Dixon, of whom all of the airmen were highly decorated and each with their own commands. H.M.A Day was Harry Day, the famous tunneller portrayed in the film *The Great Escape*.

Spending Christmas at home with his family had been a rare occurrence for Sir Brian over the years, so they enjoyed the festivities together at Bylands, their RAF house in Weybridge, Surrey. As Hogmanay passed into the new year of 1950, with only three months of his long and eventful career to serve, the family's thoughts turned to the future. One thing both Sir Brian and Jaimsie knew was that he would not be putting his feet up to retire from the busy life any time soon.

Sir Brian saw out his last three months as Commander-in-Chief at Transport Command and was keen to begin his new life. Now their first step on the way was to pack up their Weybridge house and move to Scotland, Jaimsie's childhood home and Sir Brian's' training ground – the land he loved best for many things, especially its fishing and golf. In fact, golf now took over from cricket for Sir Brian and the lure of one golf course in particular – St Andrews. So they bought a house on the edge of the Royal and Ancient at St Andrews, where Sir Brian became a member, playing nearly every day in the first few weeks. However, he knew that might not last for long … and he was right.

No sooner had it become known that Sir Brian, now almost a celebrity, was retired at 53 and living in Scotland, he became as busy as ever, with invitations to help or support a number of organisations. His first love being the RFC, the precursor of the RAF, he accepted the offer of a prestigious role to take over as the secretary and trustee of the Scottish branch of the RAF Benevolent Association, an active position he relished, starting in one month's time. Meanwhile, he found himself with more time to enjoy the occasional game of golf with old RAF friends and, more often, with his near neighbour, Dr Allen, who had served in the Medical Corps and often came round for a chat after the golf. Sir Brian's grandson Nigel recalls one day when he was staying there in the school holidays and he sat on the stairs, listening to their animated conversation:

> They were comparing notes with each other, with Grandpa being happy not to have been in the trenches and Dr Allen being happy he was not in the air. It was a very amusing conversation they had in the sitting-room, not knowing that I was quietly sitting on the stairs listening to them.

Nigel also recalls that his grandfather always wore the same tie every day. When he asked why, he learned that it was Sir Brian's Royal Flying Corps tie. Like so many of his compatriots, he had never lost his loyalty to his first allegiance, the RFC.

In between his sporting loves – golf, fishing and occasional games of cricket and tennis – Sir Brian was inundated with invitations to talk to various groups and associations and to officiate at a multitude of events, openings, anniversaries, parades and the like. One of the first of these was an address to the St Andrews Rotary Club about Transport Command. Sir Brian's host welcomed him with a brief introduction:

> No man is better qualified to speak on this subject than Sir Brian, whose name was synonymous with Transport Command. He had brought it up to its highest standard. It is rather fitting that Sir

Brian has chosen to reside at St Andrews as he was one of the best commandants ever to have presided at Leuchars.

Sir Brian had obviously been given a time limit, to which he alluded in his first sentence: 'It's a very big subject to cram into twenty minutes!' He then he took his audience back to the beginning:

In 1916 an aeroplane was used for the first time to support troops by dropping supplies. In 1920, the first Bomber Squadron dropped sacks of supplies which bumped along the ground for troops beneath.

In 1921 what was known as 'blazing of the trail' was begun by Air Marshal Sir Arthur Cunningham who, with three others, pioneered a route on the south side of the Sahara Desert. This 'blazing of the trail' extended to Iraq and to the Far East, but it was a service that was developed not by the RAF, but by Imperial Airways. Between the two wars very little was done in the way of transport aircraft in the RAF. They were told they would have to rely on civil aircraft. The result was that the Second World War began without any transport aircraft in the RAF. In 1941, when our shipping reserves were strained to the utmost, the command known as 'Ferry Command' was formed. The object of that was to ferry aircraft from America across the Atlantic.

Sir Brian listed the various countries to which these planes were ferried, before the crucial event: 'In 1942, Mr Churchill went to Africa and, on finding out that there was no mail service for the troops, he decreed that the soldiers must have their mail. A mail service to North Africa was organised.'

This, Sir Brian said, was the beginning of an RAF transport service, which was named 'Transport Command', which became indispensable to the British Army in Burma and during Operation Glasgow, the crossing of the Rhine, followed most famously by the Berlin Airlift, as Sir Brian continued to describe in the rest of his address.

Sir Brian's next talk was to the British Legion at their annual dinner, on the subject of 'government policy regarding the forces' current needs and potential further cuts'. This was a subject in which Sir Brian was well versed. He also shared his concerns on the government's plan to introduce a fifteen-day call-up, about which he voiced his very strong opinions:

> The fifteen-day call-up is quite useless. It is no use providing masses of weapons and all the technical instruments of war unless there are trained men to use them. What could be learned of modern arms in fifteen days? ... and it must be better to have a volunteer, willing to learn, than a resentful conscript.

As Sir Brian continued his talk, old soldiers of the British Legion were fully engaged and nodded frequently in agreement. He turned to his next topic: 'The lack of leadership and preparation in peacetime; Changing conditions mean changing needs'. Sir Brian brought his talk to a close with something they could all agree with – 'The enduring importance of comradeship', to warm applause.

Between the talks and other events, Sir Brian had settled into his new role as secretary of the Scottish branch of the RAF Benevolent Fund, which he took very seriously. His grandson Nigel remembers Sir Brian taking the train from St Andrews to Edinburgh two days every week, to his office, where his loyal secretary, Miss MacDougal, would have his favourite cream buns ready for him. Whenever Nigel and his brother Anthony were staying with their grandparents, Sir Brian drove them to Edinburgh with him. While he went up to his office, the boys went off to spend a few hours touring the city and the castle, which they got to know very well on those trips. When it was time to go back, they returned to Sir Brian's office, where Miss MacDougal had two cream buns waiting for them too.

From his Edinburgh office, Sir Brian organised veterans' events and fundraising opportunities. He kept in touch with all the members and their families, helping out those in need. One day in March 1968, Sir

Brian was visited by a reporter from the *Evening Citizen* newspaper, who described this encounter:

> I found Sir Brian E. Baker, KBE, CB, DSO, MC, AFC, 'forerunner of "The Few"', hunched behind a desk, strategically placed, up-sun in the Edinburgh office of the RAF Benevolent Association, of which he is Scottish Secretary. The iron-grey hair, cut short, is parted dead-centre, the chin juts, the eyes deeply set beneath thick, wiry eye-brows.
>
> As we talked about the 'first show', I could well imagine those features framed in a flying helmet, set in an expression of fierce defiance glaring through the gun-sight of a Bristol Fighter at an unfortunate member of the German 'Richthofen Circus'.
>
> At 72, the Air Marshal looks fit enough and tough enough to go ten rounds with a man half his age and as he sat, with hands clasped, almost wincing at my question about First World War heroes, I managed to extract the beginnings of the story of the Royal Flying Corps.

Next, Sir Brian was asked to become a trustee of another very important Scottish charity, the Lady MacRobert Trust. There was a very sad story at the heart of this charity, founded by Lady MacRobert, who lost all of her three sons. The eldest, Alasdair, was killed in a flying accident before the Second World War, and the other two, Roderic and Iain, joined the RAF and were killed in action during the war. This organisation is still flourishing today, funding or creating opportunities for young people to thrive and fulfil their potential.

At around this time, Sir Brian accepted an invitation involving a ride each way in a helicopter. No doubt, that would have appealed to him, but even more important was the occasion – an opportunity to see one of the benefits of his own keen fundraising over the years. When his helicopter touched down on the lawn at its destination in Erskine in Renfrewshire, the matron, Miss Cameron, stepped forward and welcomed him to the Princess Louise Scottish Hospital for disabled ex-servicemen. Sir Brian was there to open the new £50,000 canteen and social centre, all paid for by generous donations from current and former members of the forces, including Sir Brian himself. The matron and members of the St Andrews Boys' Brigade, whom Sir Brian congratulated on their excellent turnout, escorted him on his tour. He gave them a short, impromptu talk about taking pride in themselves and in everything they do, aptly illustrated by an amusing incident that happened to him when he was at school. Sir Brian admired the new facilities and spoke with some of the disabled servicemen over a cup of tea, before he left for the short flight home.

Chatting with those ex-servicemen and hearing some of their memories had given Sir Brian an idea. There must be several RFC veterans like himself and older, all over Scotland. Perhaps it would be possible to organise a gathering one day.

Meanwhile, the sun was shining at St Andrews and the verdant greens of the Royal and Ancient beckoned. Sir Brian applied to compete for the prestigious Queen Victoria Jubilee Vase and, as an R&A member with a good record on course, his entry had been accepted. So now he had to get in as much practice as possible. It was an open tournament – open to professionals and amateurs alike. When the entries closed, Sir Brian read in the newspaper the final list of confirmed competitors. His name was there, amongst the 'distinguished personalities', who included a viscount, an earl, four lords, one honorable and four Knights of the Realm, including Sir Brian himself. Grand as some of the titles were, the stiffest opposition would come from the international players, one of whom was the holder of the previous year's trophy and another was a one-armed international champion. Unfortunately, Sir Brian's sister did not paste the results into any of her albums, so presumably, Sir Brian didn't win the Jubilee Vase. However, he entered the competition to enjoy the challenge and raise his game, so he must have been happy to take his place amongst some of the world's top golfers.

In the long summer holidays, Sir Brian and Jaimsie's daughters and grandchildren came to stay. Over the decades, Jean and Margot had rarely seen their father as he had spent so much of his time in far-flung RAF postings, sometimes with Jaimsie. The girls had both gone to boarding school and spent most of their school holidays with their Aunt Winifred at the house in Hertford in which she and Sir Brian had grown up. Now, at last, his daughters and their own children could get closer to both Sir Brian and Jaimsie. The grandchildren still hold fond memories of fun times, especially helping Sir Brian to sort out and clean his fishing tackle on the lawn. Nigel's sister Jacqueline remembers the day Sir Brian filled a tub with water and helped her into his waders, then lifted her into the tub. On asking him why, Sir Brian told her he was just checking to see if they leaked.

Every year, Sir Brian took his grandsons, Nigel and Anthony, to stay at Alastrean House, the RAF holiday home in the Highlands. The river Don was close by and Sir Brian took the boys fishing there, or sometimes farther north, where he knew some of the 'lairds' and they could fish whenever they wanted. Nigel still cherishes his memories of fishing with their grandpa in the Scottish lochs and glens, either from a boat or shady riverbanks. Nigel particularly remembers his grandfather's patience in teaching him and his twin brother Anthony how to cast a fly:

From about the age of 12 we would cast the line down the garden. Grandpa showed us first, then we would cast the line ahead, down the lawn and, with practice, the fly at the end of the line gently landed ahead on the 'water', gently landing in a straight line.

It took quite a lot of practice, as if it wasn't gentle, this would frighten the fish away.

The fly rod had to be such that all the action was in the wrist. It did take quite a lot of practice, but, but only by actual fly fishing did we manage to get it right and consequently caught a trout.

My brother and I were about 14 when we first fly-fished in the river Don. Anthony didn't quite get the hang of it, so, he had the landing-net ready when Grandpa or I caught a trout.

I still think of Grandpa when I fish in my local lakes today.

At the end of a day's fishing, when they arrived back at Alastrean House and walked into the lounge bar, all the RAF men there always stood up to show their respect and gathered round Sir Brian, making him the centre of their attention. Nigel remembers how proud he and Anthony were of their grandfather.

With the young family gone and the arrival of autumn, Sir Brian returned to their lovely house in St Andrews, knowing there would be a large pile of letters from all over Scotland, plus a few from England and other countries too, most of them inviting him to open a fete or unveil a monument or give a speech or lead a parade at various armistice, peace or victory commemorations. Many of these events were to be held on the same anniversary days, so it must have been difficult to choose which invitations to accept. The most important national and international events would be considered first, especially if Sir Brian was asked to accompany Queen Elizabeth to inspect the troops, unveil a plaque or plant a tree. Having known her parents and grandparents quite well over the years, they got on very well. Other favourite criteria for choosing were if they were for veterans, such as the British Legion, which he had joined, or fund-raising events or international peace celebrations. He tried to vary them from year to year. At some stage, Sir Brian joined the RNLI, who appointed him as their governor. As ever, he was very active in this role and when he retired from it several years later, he was awarded a gold medal.

Finally, Sir Brian had gathered enough contact details to enable him to organise a grand reunion of over seventy RFC and Royal Naval Air Service veterans from the First World War. The venue was the officers' mess at RAF Turnhouse, where a good buffet supper was laid on for them all, but the most exciting and memorable aspect of the evening was undoubtedly the joy in these pioneer airmen's faces as they recognised old pals not seen for decades and the memories they eagerly shared with one another throughout the evening. Sir Brian was one of them, of course, but when he stood back and watched, he must have been proud of having been able to bring so many of them back together for one last time.

On Sunday, 15 September 1968, Sir Brian and Jaimsie attended a Service of Thanksgiving at Westminster Abbey to commemorate the victory granted in the Battle of Britain in 1940 and the fiftieth anniversary of the foundation of the Royal Air Force in 1918. A few months later, on 12 May 1969, Sir Brian and Jaimsie attended a twentieth anniversary celebration of the end of the Berlin Airlift.

By now, Sir Brian, although still playing a vigorous round of golf whenever he could, was gradually decreasing his official commitments, but there was one big day that he was determined to attend and hopefully take a role in – no matter how far he would have to travel. However, this time he was in luck. Not only were he and Jaimsie spared the journey to London, Sir Brian was delighted to be asked to help plan the special day … and to take an active part in it. Thus, on Monday, 3 July 1972, Sir Brian and Jaimsie dressed themselves up for the occasion. Sir Brian proudly did up his RFC tie and checked that he had the day's plan in his pocket, though he already knew it all by heart. They left 3 Howard Place, their St Andrews house, and drove to Edinburgh, to Holyrood House, the Scottish equivalent of Buckingham Palace. This was a familiar haunt for Sir Brian and Jaimsie over many years of attending Royal Garden Parties there, as well as investitures and other gatherings. They made their way to the correct entrance and took their places for the start of the celebration of the diamond jubilee of the formation of the Royal Flying Corps and the Royal Naval Air Service. Jaimsie remained seated while Sir Brian made his way across the parade ground to check that everyone and everything was ready. He turned and gave a nod to a royal equerry.

The programme stated: 'The parade will be commanded by Air Marshal Sir Brian Baker.' He stood upright when, on the dot, the Queen and Princess Anne arrived in the garden, ready for the presentations of various military personnel to the royal party. Sir Brian then gave the signal to the bandmaster, who conducted the playing of the National Anthem. The royal party then proceeded to the dais, whereupon Sir Brian walked across and informed the Queen that the Royal Flying Corps and the Royal Naval Air Service were ready to be inspected.

He led her to the end of the ranks of immaculately uniformed men and accompanied her as she inspected them, pausing here and there to talk with some of them. Princess Anne followed, also inspecting and talking to the men, accompanied by Air Marshal Sir Thomas Elmhirst.

Once the inspections were over, Sir Brian gave the command in his loudest voice, still very strong and clear: 'Three cheers for Her Majesty the Queen.' At this point, the Queen and all the royal party returned to the palace. Throughout the proceedings, the Band of the Royal Air Force played a medley of tunes, from 'Colonel Bogey' to 'My Fair Lady', and the members of the parade marched out to 'Heart of Oak'.

In 1978, Sir Brian, aged 82, paid tribute to his friend, the late Air Chief Marshall, Lord Dowding, by unveiling the great man's statue, for which Sir Brian had vigorously campaigned during the eight years since Dowding's death. It must have been very gratifying for Sir Brian to see him honoured at last in his own home town of Moffat. At this unveiling, Sir Brian was introduced as: 'Retired but that is a misnomer. Sir Brian has never retired. Indeed, if I tried to enumerate the list of organisations he works or fund-raises for, it would take far too long.'

On Remembrance Day in 1978, Sir Brian took part in the service at St Holy Trinity Church at St Andrews and laid a wreath. This was to be his last public appearance, for Jaimsie had died in June that year and Sir Brian himself fell ill and spent his last few weeks at Nocton Hall, the RAF nursing home in Lincolnshire, where he died, just four months after Jaimsie.

Before Sir Brian died, his young grandson, Jamie, wrote this poem:

MY GRANDPA

> My Grandpa now is eighty-three
> And sadly he is ill,
> But the service to his country
> Is well-remembered still.

In World War 1 at twenty,
He won the DSO,
The MC, and the AFC,
Made up a hero's row.

A pilot of those early planes
Which flew o'er no man's land,
He had a bird's eye view of war
Which ravaged Europe's land.

The boys with whom he went to school
Were mainly killed in France.
He himself was fortunate
That he survived by chance.

In World War two he trained and planned
Against the Nazi tyranny
Which bought dread war again.
And led much younger men

Despite his service record,
A quiet, gentle, man
Who baits a hook and casts a line
Better than most folks can.

Air Marshall Sir Brian Baker died peacefully on 9 October 1979 at Nocton Hall, aged 83. His sister, Winifred, outlived him and faithfully pasted a selection of the many condolence letters, newspaper articles and obituaries into the final pages of her last album of her beloved brother's extraordinary life, or as he might modestly have put it, 'a good innings'.

Epilogue

Shortly after Brian's funeral, a plaque was made in his memory and installed in the parish church at Hertford, his childhood home. At the unveiling ceremony, an old RAF friend of Brian's, Air Marshal Sir Charles Edward Chilton, KBE, gave a speech that summarised the man and his achievements.

This morning you will have witnessed a small interlude in which I unveiled the Memorial Plaque to honour the memory of Air Marshal Sir Brian Baker, KBE, CB, DSO, MC, AFC and *Croix de Guerre*. He was a very remarkable man in every way and, apart from his flying abilities and qualities as an officer in the Royal Air Force, he was an outstanding sportsman in many different fields. He was a very fine shot, an excellent fisherman and an outstanding cricketer and, above all, he was a fine gentleman and was loved by all.

In my own 38 years in the RAF, I have never heard one word about Sir Brian which could be said to be other than admiration and affection. I can say that my own life has been greatly enhanced by my friendship with Brian. Over the years, he gave me much wise advice and inspiration. Indeed, I came to know him so well that I almost hear him whispering to me now: 'If you must speak, make it short.' He did not approve of showmanship.

'E.B.E.', as he was known to all, was born in 1896, attended this church, where he was christened, until he retired from the RAF in 1959. He always gave his family home here as his address as he loved the area and it closely linked him with his sister Winnie, to whom he was devoted. It is her inspiration, with the willing help of your Vicar, which has brought this plaque into being.

Going back in time a little, Brian left his public school, Haileybury, as soon as he could to join the war. He very quickly made his mark and was soon 'mentioned in dispatches' and then went on to win the DSO, MC, AFC and the *Croix de Guerre*. After just a year, he was to become the Chief Flying Instructor at Leuchars in Scotland. It was this appointment which was to fashion the rest of his career, as he was to be closely linked with the maritime side of the RAF for so many years.

As AOC of Number 19 group he was responsible for keeping the German submarines out of the invasion area for D-Day and it was 100% successful – a remarkable story, almost unknown to the British public.

He also played a great part in the Berlin Airlift, another important event that we all seem to have forgotten. They were probably the most onerous posts of any service.

Brian married Jaimsie Robinson, who sadly died a short time before him. They were probably the most generous hosts in the service. Everyone was welcome at their house and during the cricket season the house was filled to capacity.

When Brian retired, it was typical that he elected to take over the affairs of the RAF Benevolent Society in Scotland. He put his heart and soul into this very worthy project; he saw to it that every case had the most personal and understanding treatment.

To close, I ask you to remember Brian as essentially a kind, honourable and generous man, who served his country well. He is greatly missed by his sister, his daughters and his grandchildren … and his many friends.

At the beginning of this book, as a 13-year-old schoolboy at Hendon Air Show, Brian overheard an official-looking man remarking: 'This is all very well, but it's madness to think that these flying machines will ever be any use.' Brian had been so sure the man was wrong that, with the certainty of youth, he made up his mind to prove it – and that is just what he did, most emphatically, throughout his flying life.

Acknowledgements

Firstly, I would like to thank the late Winifred Baker, without whose extensive archives of her brother's life and achievements this book could not have been written.

Nigel Butler for his great help, support and memories of his grandfather.

Margot Butler for additional material and her memories of her father.

Malcolm Fleming, Margo's cousin, who shared his memories of Sir Brian.

Toby Parker, Haileybury Archivist, for his helpful information and extracts from several school magazines.

Richard Brewis at Montrose Heritage Centre for his helpful correspondence.

Hertford Museum for access to First World War transcripts.

Jill Wadsworth at Stevenage Museum and Alan Ford for providing information and the photo of a pre-First World War air crash.

Bibliography

Books

Ashcroft, R.L., *Haileybury, 1908–1961*, Haileybury (Hertford, 1961)

Barker, Ralph, *The Royal Flying Corps in World War I*, Constable & Robinson (London, 2002)

Levine, Joshua, *On a Wing and a Prayer*, Collins (London, 2008)

McCudden, James Byford, *Flying Fury*, Casemate (Newbury, 1983)

Pitchfork, Graham, *The Royal Air Force Day by Day*, Sutton Publishing (Stroud, 2008)

Shores, Christopher, Franks, Norman & Guest, Russell, *Above the Trenches*, Grub Street (London, 1937)

Thomas, J.B.W., *Sursum Corda*, Haileybury Centenary Number MCMLXII

Wallace, Graham, *R.A.F. Biggin Hill*, Putnam (London, 1957)

Wilson, Michael & Robinson, A.S.L, *Coastal Command Leads the Invasion* (London, date unknown)

Digital sources

Bournemouth University

D-Day Revisited

Combined Ops

Cross and Cockade

Fairfields Airfields

Forces of War

Forces WAR Records UK

Friends of Stokes Bay

History.com

Imperial War Museum

KUMC.Education

Medium

Military Wiki

Mischief Night – Allied Radar

Montrose Air Station Heritage Centre

Ontario Records

Radar Museum

RAF Web
Smithsonian Institution
South Dublin Libraries
The Aerodrome
The Genealogist RAF Operations Record Books
The Scholarly Community Encyclopaedia
The Wartime Memories Project
Traces at War
Trustees of the RAF Museum
UK History
Valley Aviation Society
Voices of the First World War in the Air
Wikipedia
Wired.com
100 Schools Resources

Letters, documents and ephemera
'Auntie Winnie' amassed a vast private collection, which included:
Sir Brian's handwritten 1915 letters while flight training
Many more of Sir Brian's handwritten letters from war zones
Pilots' log books
Citations, decorations, medals and honours from Britain and various other nations
Texts of Sir Brian's speeches, talks, lectures, articles and broadcasts
Original documents such as orders, an aerial map, messages from heads of nations
    to stir or congratulate officers and men
Personal telegrams from illustrious names
Royal invitations to investitures, garden parties, private parties, national events etc.
King Farouk's palace party invitation and programme
Records of Sir Brian's sporting achievements (cricket, golf, tennis, rugby and
    fishing)
Auntie Winnie's photo albums and many other photos
Drawings and cartoons of Sir Brian
Original newspapers (see separate list)
Correspondence relating to Sir Brian's post-retirement voluntary work
Letters of condolences and tributes following Sir Brian's death.

Newspapers and periodicals
(Featuring Sir Brian and/or articles he wrote)
Courier and Advertiser
Daily Mirror
Daily Sketch

Dundee Courier
Eastleigh Evening News
Edinburgh Evening News
Evening Citizen
Hertford Borough & Hertford Rural
London Gazette
The People's Journal
The Scotsman
The Times
Yorkshire Mercury
Yorkshire Post

And several others, cropped by Auntie Winnie, of their headers in her albums.

Index

Dear Reader,

We hope you have enjoyed this book, but why not share your views on social media? You can also follow our pages to see more about our other products: facebook.com/penandswordbooks or follow us on X @penswordbooks

You can also view our products at www.pen-and-sword.co.uk (UK and ROW) or www.penandswordbooks.com (North America).

To keep up to date with our latest releases and online catalogues, please sign up to our newsletter at: www.pen-and-sword.co.uk/newsletter

If you would like a printed catalogue with our latest books, then please email: enquiries@pen-and-sword.co.uk or telephone: 01226 734555 (UK and ROW) or email: uspen-and-sword@casematepublishers.com or telephone: (610) 853-9131 (North America).

We respect your privacy and we will only use personal information to send you information about our products.

Thank you!